FUNDAMENTALS OF RF CIRCUIT DESIGN

with Low Noise Oscillators

FUNDAMENTALS OF RF CIRCUIT DESIGN

with Low Noise Oscillators

Jeremy Everard

University of York, UK

JOHN WILEY & SONS, LTD

Chichester • New York • Weinheim • Brisbane • Singapore • Toronto

Other Wiley Editorial Offices

John Wiley & Sons, Inc., 605 Third Avenue,
New York, NY 10158-0012, USA

Wiley-VCH Verlag GmbH
Pappelallee 3, D-69469 Weinheim, Germany

Jacaranda Wiley Ltd, 33 Park Road, Milton,
Queensland 4064, Australia

John Wiley & Sons (Canada) Ltd, 22 Worcester Road
Rexdale, Ontario, M9W 1L1, Canada

John Wiley & Sons (Asia) Pte Ltd, 2 Clementi Loop #02-01,
Jin Xing Distripark, Singapore 129809

British Library Cataloguing in Publication Data

A catalogue record for this book is available from the British Library

ISBN 0 471 49793 2

Produced from Word files supplied by the authors

This book is printed on acid-free paper responsibly manufactured from sustainable forestry,
in which at least two trees are planted for each one used for paper production.

To my wife Sue and children James, Katherine and Sarah for the lost hours and to my parents for their unquestioning support.

Contents

Preface

The number of telecommunications systems is expanding at an ever increasing rate, to the extent that most people now carry or are regularly influenced by such items. These include, for example, mobile telephones, personal stereos, radio pagers, televisions and of course the associated test equipment. The expansion of wireless local area networks and multimedia transmission by microwaves is likely to further fuel this increase. Modern computer systems also clock at RF/microwave signal rates requiring high frequency design techniques. The borderline between RF and microwave systems is also less obvious in that most, if not all, of the techniques described in this book can also be applied at microwave frequencies. For example, it is now possible to obtain low cost packaged silicon devices with an f_T greater than 65GHz. The skill required by the engineers working in this field is very broad and therefore an in-depth understanding of the underlying/fundamental principles used is very important.

The aim of this book is to explain the fundamentals of the basic building blocks used in RF circuit design both at the component and intermediate block level. At block level this includes low noise small signal amplifiers, both narrowband and broadband, low phase noise oscillators, mixers and power amplifiers. The components include bipolar transistors, FETs, resistors, capacitors, inductors, varactor diodes and diode detectors. Charts of performance parameters for chip components are included.

The approach is both theoretical and practical explaining the principles of operation and then applying theory (largely algebra) to show how significant insight, both linear and non-linear, can be obtained by using simplifications and approximations. Where necessary more accurate models can be derived by incorporating second order effects. The mathematics is generally included in full as it is important, when extensive CAD is used, that the initial analysis should use sufficient theory to show the required insight. This then enables more robust and longer lasting designs.

The book is an extension of the course material provided to delegates on advanced one-week intensive courses offered to industry by the University of York. These are offered either as a single course or as part of the Integrated Graduate Development Scheme (IGDS) masters degree programmes initially sponsored by the UK Engineering and Physical Sciences Research Council. This material is now also presented to our fourth year MEng students in the RF and Microwave Circuit Design course as part of the UK Radio Frequency Engineering Education Initiative (RFEEI). The material is presented over 27 lectures and three laboratory classes.

The book is based on both research and teaching material. The chapter on low phase noise oscillators is based on research carried out by my research group over the last 18 years. This chapter reviews oscillator design techniques and describes the latest techniques and publications to September 2000. As in the other chapters simple algebra is used to quantify most of the important parameters in low noise oscillator design to a high degree of accuracy. These include such parameters as the optimum coupling coefficients of the resonator to the amplifier and the noise caused by the varactor diode. This theory is then illustrated in eight designs showing accurate prediction of noise performance to within 0 to 2dB of the theory. The latest addition includes a new technique to remove flicker noise in microwave oscillators.

Chapter 1 describes models for bipolar transistors, FETs, varactor diodes, diode detectors and passive components including resistors capacitors and inductors. Modelling of the bipolar transistor starts with the simple T model, which most closely resembles the actual device. The π model is then developed. Both T and π models are used for the bipolar transistor, as it is often easier to solve a problem by using their equivalence and switching between them during calculations. They also offer different insights for different circuit configurations. The Miller effect is then described generically enabling the standard approximation for roll-off in bipolar transistors caused by the feedback capacitance. This is then extended to model the effect of any feedback impedance. This is later used in Chapter 3 for broadband amplifier design with optimum input and output match over multiple octave bandwidths when the feedback capacitor is replaced by a resistor.

S_{21} vs frequency and current is then derived for the π model using only the operating current, f_T and feedback capacitance. It is then shown how this can be made more accurate by incorporating the base spreading resistance and the emitter contact resistance and how this is then usually within a few per cent of the parameters given in data sheets. The accuracy is illustrated graphically for two devices operating at 1 and 10mA. This then allows intuitive design enabling the important parameters for the device to be chosen in advance, through a deeper understanding of their operation, without relying on data sheets. The harmonic and third order intermodulation distortion is then derived for common emitter and differential amplifiers showing the removal of even order terms during differential operation. The requirement for low level operation for low distortion is then illustrated in tabular form. The characteristics for FETs and varactor diodes are then described.

The operation of diode detectors is then described with a calculation of the sensitivity and illustration of the changeover between the square law and linear characteristic. The noise performance is then illustrated using the Tangential Sensitivity. Models including the parasitics and hence frequency response of chip resistors and capacitors are then described illustrating for example the effect of series resonance in capacitors and the change in impedance for resistors. This is similarly applied to inductors where empirical equations are quoted for inductors, both wound and spiral. The calculation of inductance for a toroid is then derived from first principles using Ampère's law illustrating how easy and accurate a simple fundamental calculation can be.

In **Chapter 2**, two port parameter definitions (h, z, y and S) are shown illustrating the common nature of these parameters and how a knowledge of these enables the different elements of equivalent circuit models to be deduced. Here, parameter conversion is used, for example, to deconvolve the non-linear capacitors within the device model, enabling the development of large signal models for power amplifiers. This is also used for linear models elsewhere. Transmission line characteristics are then illustrated and a simplified model for S parameters is derived which enables easier calculation of the forward and reverse parameters. This is then applied to a range of circuits including the bipolar transistor described elsewhere in the text.

Chapter 3 describes small signal amplifier design using both Y and S parameters illustrating how both approaches offer further insight. The simple calculation of the resistance required to maintain stability is illustrated using the simple S parameter equations for the input and output reflection coefficient. These same equations are later used to illustrate and calculate the models for one port error correction. Matching is described using both tapped resonant networks and two component inductor/capacitor networks using Smith Charts with a number of design examples. The effect of loaded and unloaded Q on insertion loss and hence noise figure is described.

Noise measurement and calculation are described using a two temperature technique. This is a fundamental technique which is similar in concept to commercial systems and can easily be built in-house. The bipolar transistor models are extended using the concept of complex current gain to illustrate how low noise can be obtained at the same time as optimum match by using an emitter inductor. This is similar to the method described by Hayward.

Broadband amplifier design is described in detail showing the effect of the feedback resistance, the emitter resistance and the bias current. A design example is included.

Methods for passive and active biasing of devices are then discussed. Measurements illustrating device test jigs and the operation of a modern network analyser are described. The importance of calibration and hence error correction is applied with the detailed equations for one port error correction.

Chapter 4 describes to a large extent a linear theory for low noise oscillators and shows which parameters explicitly affect the noise performance. From these analyses equations are produced which accurately describe oscillator performance usually to within 0 to 2dB of the theory. It shows that there are optimum coupling coefficients between the resonator and the amplifier to obtain low noise and that this optimum is dependent on the definitions of the oscillator parameters. The factors covered are: the noise figure (and also source impedance seen by the amplifier); the unloaded Q; the resonator coupling coefficient and hence Q_L/Q_o and closed loop gain; the effect of coupling power out of the oscillator; the loop amplifier input and output impedances and definitions of power in the oscillator; tuning effects including the varactor Q and loss resistance, the coupling coefficient of the varactor; and the open loop phase shift error prior to loop closure.

Optimisation of parameters using a linear analytical theory is of course much easier than using non-linear theories.

The chapter then includes eight design examples which use inductor/capacitor, Surface Acoustic Wave (SAW), transmission line, helical and dielectric resonators at 100MHz, 262MHz, 900MHz, 1800MHz and 7.6GHz. These oscillator designs show very close correlation with the theory usually within 2dB of the predicted minimum. It also includes a detailed design example.

The chapter then goes on to describe the four techniques currently available for flicker noise measurement and reduction including the latest techniques developed by my research group in September 2000. Here a feedforward amplifier is used to suppress the flicker noise in a microwave GaAs based oscillator by 20dB. The theory in this chapter accurately describes the noise performance of this oscillator, within the thermal noise regime, to within ½ to 1dB of the predicted minimum.

A brief introduction to a method for breaking the loop at any point, thus enabling non-linear computer aided analysis of oscillating (autonomous) systems, is described. This enables prediction of the biasing, output power and harmonic spectrum.

Chapter 5 describes an introduction to mixers starting with a simple non-linear device and then leading on to an ideal switching mixer. The operation and waveforms of switching diode and transistor mixers are then described. Diode parameters such as gain compression and third order intermodulation distortion are then introduced.

Chapter 6 provides an introduction to power amplifier design and includes Load Pull measurement and design techniques and a more analytic design example of a broadband, efficient amplifier operating from 130 to 180 MHz. The design example is based around high efficiency Class E techniques and includes the development of an accurate large signal model for the active device. This model is also used to enable calculation of the large signal input impedance of the device under the correct operating conditions. Although the design relates to Class E techniques the methods described can be used for all amplifier types.

Chapter 7 describes a circuit simulator which displays in real time the waveforms, at all the nodes, while using the mouse with crosshatch and slider controls to vary the component values and frequency at the same time as solving the relevant differential equations. This then enables real time tuning of the circuit for optimum response. The techniques for entering the differential equations for the circuit are described. These differential equations are computed in difference form and are calculated sequentially and repetitively while the component values and frequency are varied. This is similar to most commercial time domain simulators, but it is shown here that it is relatively easy to write down the equations for fairly simple circuits. This also provides insight into the operation of large signal simulators. The version presented here uses Visual Basic Version 6 for a PC and enables the data to be presented in an easily readable format. A version of this program is used here to examine the response of a broadband highly efficient amplifier load network operating around 1 to 2GHz.

Summary: The aim therefore is to provide a book which contains both analytical and practical information enabling insight and advanced design through in-depth understanding of the important parameters.

I plan to maintain a web page with addenda, corrections and answers to any comments by readers. The URL will be a subfolder of my University of York web page: http://www.york.ac.uk/depts/elec/staff/academic/jkae.html.

Acknowledgements

I would like to thank: Rob Sloan for enormous help with the figures and diagrams, Carl Broomfield for many technical discussions, proof reading, help with experiments and help with many of the graphs and Pete Turner for help in writing the Visual Basic simulator on 'real time' circuit modelling in Chapter 7.

I would also like to thank Paul Moore from Philips Research Laboratories who, in the early 1980s, started me in the right direction on oscillator design. I also thank Jens Bitterling, Michael Cheng, Fraser Curley, Paul Dallas, Michael Page-Jones and Andrew King, all former members of my research groups, for their help in generating new research results. I would like to thank the UK Engineering and Physical Sciences Research Council for supporting most of the research work on oscillators and for their support of the IGDS MSc program.

Finally I would like to thank the University of York and also my colleagues for helpful discussions. I would also like to thank Peter Mitchell, Kathryn Sharples and Robert Hambrook from Wiley for their help and patience.

Jeremy K.A.Everard
University of York, UK
jkae@ohm.york.ac.uk

1

Transistor and Component Models at Low and High Frequencies

1.1   Introduction

Equivalent circuit device models are critical for the accurate design and modelling of RF components including transistors, diodes, resistors, capacitors and inductors. This chapter will begin with the bipolar transistor starting with the basic T and then the π model at low frequencies and then show how this can be extended for use at high frequencies. These models should be as simple as possible to enable a clear understanding of the operation of the circuit and allow easy analysis. They should then be extendible to include the parasitic components to enable accurate optimisation. Note that knowledge of both the T and π models enables regular switching between them for easier circuit manipulation. It also offers improved insight.

As an example S_{21} for a bipolar transistor, with an f_T of 5GHz, will be calculated and compared with the data sheet values at quiescent currents of 1 and 10mA. The effect of incorporating additional components such as the base spreading resistance and the emitter contact resistance will be shown demonstrating accuracies of a few per cent.

The harmonic and third order intermodulation distortion will then be derived for common emitter and differential amplifiers showing the removal of even order terms during differential operation.

The chapter will then describe FETs, diode detectors, varactor diodes and passive components illustrating the effects of parisitics in chip components.

It should be noted that this chapter will use certain parameter definitions which will be explained as we progress. The full definitions will be shown in Chapter 2. Techniques for equivalent circuit component extraction are also included in Chapter 2.

1.2   Transistor Models at Low Frequencies

1.2.1   'T' Model

Considerable insight can be gained by starting with the simplest T model as it most closely resembles the actual device as shown in Figure 1.1. Starting from a basic NPN transistor structure with a narrow base region, Figure 1.1a, the first step is to go to the model where the base emitter junction is replaced with a forward biased diode.

The emitter current is set by the base emitter junction voltage The base collector junction current source is effectively in parallel with a reverse biased diode and this diode is therefore ignored for this simple model. Due to the thin base region, the collector current tracks the emitter current, differing only by the base current, where it will be assumed that the current gain, β, remains effectively constant.

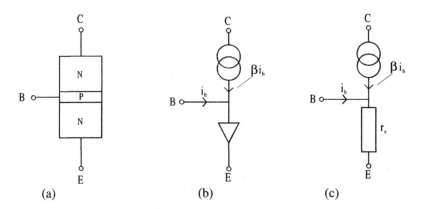

Figure 1.1 Low frequency 'T' model for a bipolar transistor

Note that considerable insight into the large signal behaviour of bipolar transistors can be obtained from the simple non-linear model in Figure 1.1b. This will be used later to demonstrate the harmonic and third order intermodulation distortion in a common emitter and differential amplifier. Here, however, we will concentrate on the low frequency small signal AC 'T' model which takes into account the DC bias current, which is shown in Figure 1.1c. Here r_e is the AC resistance of the forward biased base emitter junction.

The transistor is therefore modelled by an emitter resistor r_e and a controlled current source. If a base current, i_b, is applied to the base of the device a collector current of βi_b flows through the collector current source. The emitter current, I_E, is therefore $(1+\beta)i_b$. The AC resistance of r_e is obtained from the differential of the diode equation. The diode equation is:

$$I_E = I_{ES}\left(\exp\left(\frac{eV}{kT}\right) - 1\right) \tag{1.1}$$

where I_{ES} is the emitter saturation current which is typically around 10^{-13}, e is the charge on the electron, V is the base emitter voltage, V_{be}, k is Boltzmann's constant and T is the temperature in Kelvin. Some authors define the emitter current, I_E, as the collector current I_C. This just depends on the approximation applied to the original model and makes very little difference to the calculations. Throughout this book equation (1) will be used to define the emitter current.

Note that the minus one in equation (1.1) can be ignored as I_{ES} is so small. The AC admittance of r_e is therefore:

$$\frac{dI}{dV} = \frac{e}{kT} I_{ES} \exp\left(\frac{eV}{kT}\right) \tag{1.2}$$

Therefore:

$$\frac{dI}{dV} = \frac{e}{kT} I \tag{1.3}$$

The AC impedance is therefore:

$$\frac{dV}{dI} = \frac{kT}{e} \cdot \frac{1}{I} \tag{1.4}$$

As $k = 1.38 \times 10^{-23}$, T is room temperature (around 20°C) = 293K and e is 1.6×10^{-19} then:

$$r_e = \frac{dV}{dI} \approx \frac{25}{I_{mA}}$$

(1.5)

This means that the AC resistance of r_e is inversely proportional to the emitter current. This is a very useful formula and should therefore be committed to memory. The value of r_e for some typical values of currents is therefore:

$1\text{mA} \approx 25\Omega$
$10\text{mA} \approx 2.5\Omega$
$25\text{mA} \approx 1\Omega$

It would now be useful to calculate the voltage gain and the input impedance of the transistor at low frequencies and then introduce the more common π model. If we take a common emitter amplifier as shown in Figure 1.2 then the input voltage across the base emitter is:

$$V_{in} = (\beta + 1)i_b . r_e$$

(1.6)

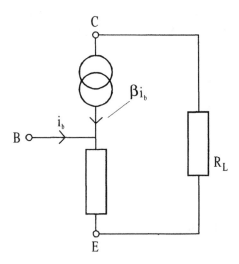

Figure 1.2 A common emitter amplifier

The input impedance is therefore:

$$Z_{in} = \frac{V_{in}}{i_b} = \frac{i_b(1+\beta)r_e}{i_b} = (1+\beta)r_e = (1+\beta)\frac{25}{I_{mA}} \qquad (1.7)$$

The forward transconductance, g_m, is:

$$g_m = \frac{I_{out}}{V_{in}} = \frac{\beta i_b}{i_b(\beta+1)r_e} \approx \frac{1}{r_e} \qquad (1.8)$$

Therefore:

$$g_m \approx \frac{1}{r_e} \qquad (1.9)$$

and:

$$\frac{V_{out}}{V_{in}} = -g_m R_L = -\frac{R_L}{r_e} \qquad (1.10)$$

Note that the negative sign is due to the signal inversion.

Thus the voltage gain increases with current and is therefore equal to the ratio of load impedance to r_e. Note also that the input impedance increases with current gain and decreases with increasing current.

In common emitter amplifiers, an external emitter resistor, R_e, is often added to apply negative feedback. The voltage gain would then become:

$$\frac{V_{out}}{V_{in}} = \frac{R_L}{r_e + R_e} \qquad (1.11)$$

Note also that part or all of this external emitter resistor is often decoupled and this part would then not affect the AC gain but allows the biasing voltage and current to be set more accurately. For the higher RF/microwave frequencies it is often preferable to ground the emitter directly and this is discussed at the end of Chapter 3 under DC biasing.

1.2.2   The π Transistor Model

The 'T' model can now be transformed to the π model as shown in Figure 1.3. In the π model, which is a fully equivalent and therefore interchangeable circuit, the input impedance is now shown as $(\beta+1)r_e$ and the output current source remains the same. Another format for the π model could represent the current source as a voltage controlled current source of value $g_m V_1$. The input resistance is often called *r*π.

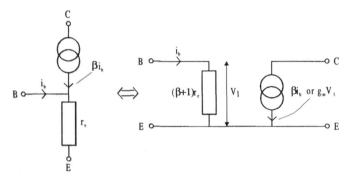

Figure 1.3 T to π model transformation

At this point the base spreading resistance $r_{bb'}$ should be included as this incorporates the resistance of the long thin base region. This typically ranges from around 10 to 100Ω for low power discrete devices. The node interconnecting *r*π and $r_{bb'}$ is called *b'*.

1.3   Models at High Frequencies

As the frequency of operation increases the model should include the reactances of both the internal device and the package as well as including charge storage and transit time effects. Over the RF range these aspects can be modelled effectively using resistors, capacitors and inductors. The hybrid π transistor model was therefore developed as shown in Figure 1.4. The forward biased base emitter junction and the reverse biased collector base junction both have capacitances and these are added to the model. The major components here are therefore the input capacitance $C_{b'e}$ or C_π and the feedback capacitance $C_{b'c}$ or C_μ. Both sets of symbols are used as both appear in data sheets and books.

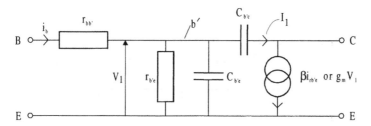

Figure 1.4 Hybrid π model

A more complete model including the package characteristics is shown in Figure 1.5. The typical package model parameters for a SOT 143 package is shown in Figure 1.6. It is, however, rather difficult to analyse the full model shown in Figures 1.5 and 1.6 although these types of model are very useful for computer aided optimisation.

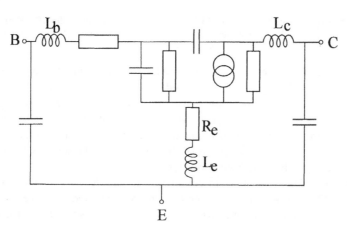

Figure 1.5 Hybrid π model including package components

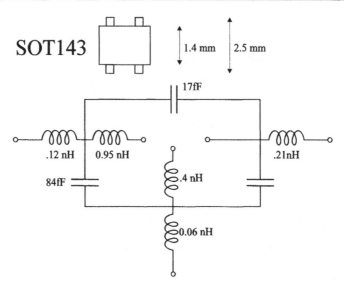

Figure 1.6. Typical model for the SOT143 package. Obtained from the SPICE model for a BFG505. Data in Philips RF Wideband Transistors CD, Product Selection 2000 Discrete Semiconductors.

We should therefore revert to the model for the internal active device for analysis, as shown in Figure 1.4, and introduce some figures of merit for the device such as f_β and f_T. It will be shown that these figures of merit offer significant information but ignore other aspects. It is actually rather difficult to find single figures of merit which accurately quantify performance and therefore many are used in RF and microwave design work. However, it will be shown later how the S parameters can be obtained from knowledge of f_T.

It is worth calculating the short circuit current gain h_{21} for this model shown in Figure 1.4. The full definitions for the h, y and S parameters are given in Chapter 2. h_{21} is the ratio of the current flowing out of port 2 into a **short circuit load** to the input current into port 1.

$$h_{21} = \frac{I_c}{I_b} \qquad (1.12)$$

The proportion of base current, i_b, flowing into the base resistance, $r_{b'e}$, is therefore:

$$i_{rb'e} = \frac{\dfrac{1}{r_{b'e}} \cdot i_b}{j\omega(C_{b'e} + C_{b'c}) + \dfrac{1}{r_{b'e}}} = \frac{i_b}{j\omega CR + 1} \tag{1.13}$$

where the input and feedback capacitors add in parallel to produce C and the $r_{b'e}$ becomes R. The collector current is $I_c = \beta\, i_{rb'e}$, where we assume that the current through the feedback capacitor can be neglected as $I_{Cb'c} << \beta i_{rb'e}$. Therefore:

$$h_{21} = \frac{I_c}{i_b} = \frac{\beta}{SCR + 1} = \frac{h_{fe}}{SCR + 1} \tag{1.14}$$

Note that β and h_{fe} are both symbols used to describe the low frequency current gain.

A plot of h_{21} versus frequency is shown in Figure 1.7. Here it can be seen that the gain is constant and then rolls off at 6dB per octave. The transition frequency f_T occurs when the modulus of the short circuit current gain is 1. Also shown on the graph, is a trace of h_{21} that would be measured in a typical device. This change in response is caused by the other parasitic elements in the device and package. f_T is therefore obtained by measuring h_{21} at a frequency of around $f_T/10$ and then extrapolating the curve to the unity gain point. The frequency from which this extrapolation occurs is usually given in data sheets.

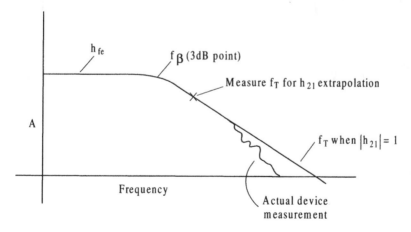

Figure 1.7 Plot of h_{21} vs frequency

The 3 dB point occurs when $\omega CR = 1$. Therefore:

$$f_\beta = \frac{1}{2\pi CR} \qquad CR = \frac{1}{2\pi f_\beta} \tag{1.15}$$

and h_{21} can also be expressed as:

$$h_{21} = \frac{h_{fe}}{1 + j\dfrac{f}{f_\beta}} \tag{1.16}$$

As f_T is defined as the point at which $|h_{21}| = 1$, then:

$$\left| \frac{h_{fe}}{1 + j\dfrac{f_T}{f_\beta}} \right| = 1 = \frac{h_{fe}}{\sqrt{1 + \left(\dfrac{f_T}{f_\beta}\right)^2}} \tag{1.17}$$

$$1 + \left(\frac{f_T}{f_\beta}\right)^2 = \left(h_{fe}\right)^2 \tag{1.18}$$

$$\left(\frac{f_T}{f_\beta}\right)^2 = \left(h_{fe}\right)^2 - 1 \tag{1.19}$$

As:

$$\left(h_{fe}\right)^2 \gg 1 \tag{1.20}$$

$$f_T = h_{fe}\cdot f_\beta = \frac{h_{fe}}{2\pi CR} \tag{1.21}$$

Note also that:

$$f_\beta = \frac{f_T}{h_{fe}} \qquad (1.22)$$

As:

$$h_{21} = \frac{h_{fe}}{1 + j\dfrac{f}{f_\beta}} \qquad (1.16)$$

it can also be expressed in terms of f_T :

$$h_{21} = \frac{h_{fe}}{1 + j\dfrac{h_{fe} \cdot f}{f_T}} \qquad (1.23)$$

Take a typical example of a modern RF transistor with the following parameters: f_T = 5 GHz and h_{fe} = 100. The 3dB point for h_{21} when placed directly in a common emitter circuit is f_β = 50MHz.

Further information can also be gained from knowledge of the operating current. For example, in many devices, the maximum value of f_T occurs at currents of around 10mA. For these devices (still assuming the same f_T and h_{fe}) r_e = 2.5Ω, therefore $r_{b'e}$ ≈ 250Ω and hence C_T ≈ 10pF with the feedback component of this being around 0.5 to 1pF.

For lower current devices operating at 1mA (typical for the BFT25) r_e is now around 25Ω, $r_{b'e}$ around 2,500Ω and therefore C_T is a few pF with $C_{b'e}$ ≈ 0.2pF.

Note, in fact, that these calculations for C_T are actually almost independent of h_{fe} and only dependent on I_C, r_e or g_m as the calculations can be done in a different way. For example:

$$CR = 1/2\pi f_\beta = h_{fe}/2\pi f_T \qquad (1.24)$$

Therefore:

$$\dot{C} = \frac{h_{fe}}{2\pi f_T \left(h_{fe}+1\right)r_e} \approx \frac{1}{2\pi f_T r_e} = \frac{g_m}{2\pi f_T} \tag{1.25}$$

Many of the parameters of a modern device can therefore be deduced just from f_T, h_{fe}, I_c and the feedback capacitance with the use of these fairly simple models.

1.3.1   Miller Effect

f_T is a commonly used figure of merit and is quoted in most data sheets. It is now worth discussing f_T in detail to find out what other information is available.

1. What does it hide?   Any output components as there is a short circuit on the output.

2. What does it ignore?   The effects of the load impedance and in particular the Miller effect. (It does include the effect of the feedback capacitor but only into a short circuit load.)

It is important therefore to investigate the effect of the feedback capacitor when a load resistance R_L is placed at the output. Initially we will introduce a further simple model.

If we take the simple model shown in Figure 1.8, which consists of an inverting voltage amplifier with a capacitive feedback network, then this can be identically modelled as a voltage amplifier with a larger input capacitor as shown in Figure 1.8b. The effect on the output can be ignored, in this case, because the amplifier has zero output impedance.

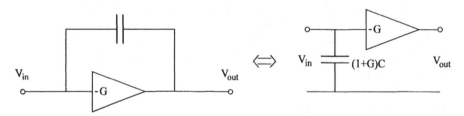

Figure 1.8a Amplifier with feedback C   **Figure 1.8.b** Amplifier with increased input C

This is most easily understood by calculating the voltage across the capacitor and hence the current flowing into it. The voltage across the feedback capacitor is:

$$V_c = \left(V_{in} - V_{out}\right)$$ (1.26)

If the voltage gain of the amplifier is $-G$ then the voltage across the capacitor is therefore:

$$V_c = \left(V_{in} + GV_{in}\right) = V_{in}\left(1 + G\right)$$ (1.27)

The current through the capacitor, I_c, is therefore $I_c = V_c j\omega C$. The change in input admittance caused by this capacitor is therefore:

$$\frac{I_c}{V_{in}} = \frac{V_c j\omega C}{V_{in}} = \frac{V_{in}\left(1 + G\right)j\omega C}{V_{in}} = \left(1 + G\right)j\omega C$$ (1.28)

The capacitor in the feedback circuit can therefore be replaced by an input capacitor of value $(1 + G)C$. This is most easily illustrated with an example. Suppose a 1V sinewave was applied to the input of an amplifier with an inverting gain of 5. The output voltage would swing to -5V when the input was $+1$V therefore producing 6V $(1 + G)$ across the capacitor. The current flowing into the capacitor is therefore six times higher than it would be if the same capacitor was on the input. The capacitor can therefore be transferred to the input by making it six times larger.

1.3.2   Generalised 'Miller Effect'

Note that it is worth generalising the 'Miller effect' by replacing the feedback component by an arbitrary impedance Z as shown in Figure 1.8c and then investigating the effect of making Z a resistor or inductor. This will also be useful when looking at broadband amplifiers in Chapter 3 where the feedback resistor can be used to set both the input and output impedance as well as the gain. It is also worth investigating the effect of changing the sign of the gain.

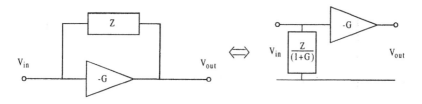

Figure 1.8c Generalised Miller effect   **Figure 1.8d** Generalised Miller effect

As before:

$$V_Z = (1+G)V_{in} \tag{1.29}$$

As:

$$I_Z = \frac{V_Z}{Z} \tag{1.30}$$

the new input impedance is now:

$$\frac{V_{in}}{I_Z} = \frac{Z}{(1+G)} \tag{1.31}$$

as shown in Figure 1.8d. If Z is now a resistor, R, then the input impedance becomes:

$$\frac{R}{(1+G)} \tag{1.32}$$

This information will be used later when discussing the design of broadband amplifiers.

If Z is an inductor, L, then the input impedance becomes:

$$\frac{j\omega L}{(1+G)} \tag{1.33}$$

As before, if Z was a capacitor then the input impedance becomes:

$$\frac{-j}{\omega C(1+G)} \tag{1.34}$$

If the gain is set to be positive and the feedback impedance is a resistor then the input impedance would be:

$$\frac{R}{(1-G)} \tag{1.35}$$

which produces a negative resistance when G is greater than one.

1.3.3   Hybrid π Model

It is now worth applying the Miller effect to the hybrid π model where for convenience we make the current source voltage dependent as shown in Figure 1.9.

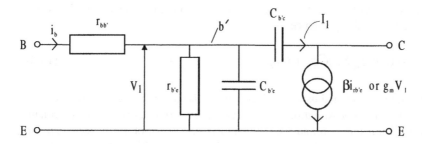

Figure 1.9 Hybrid π model for calculation of Miller capacitance

Firstly apply the Miller technique to this model. As before it is necessary to calculate the input impedance caused by $C_{b'c}$. The current flowing into the collector load, R_L, is:

$$I_c = g_m V_1 + I_1 \tag{1.36}$$

where the current through $C_{b'c}$ is:

$$I_1 = (V_1 - V_0)C_{b'c}.j\omega \tag{1.37}$$

The feedforward current I_1 through the feedback capacitor $C_{b'c}$ is usually small compared to the current $g_m V_1$ and therefore:

$$V_0 \approx -g_m R_L V_1 \tag{1.38}$$

Therefore:

$$I_1 = (V_1 + g_m R_L V_1) C_{b'c} \cdot j\omega \qquad (1.39)$$

$$I_1 = V_1 (1 + g_m R_L) C_{b'c} \cdot j\omega \qquad (1.40)$$

The input admittance caused by $C_{b'c}$ is:

$$\frac{I_1}{V_1} = (1 + g_m R_L) C_{b'c} \cdot j\omega \qquad (1.41)$$

which is equivalent to replacing $C_{b'c}$ with a shunt capacitor in parallel with $C_{b'e}$ to produce a total capacitance:

$$C_T = (1 + g_m R_L) C_{b'c} + C_{b'e} \qquad (1.42)$$

The model generated using the Miller effect is shown in Figure 1.10. Note that this model is an approximation in this case, as it is only effective for calculating the forward transmission and the input impedance. It is not useful for calculating the output impedance or the reverse transmission or stability. This is only because of the approximation used when deriving the output voltage. If the load is zero then the current gain would be as derived when h_{21} was calculated.

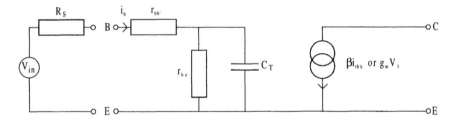

Figure 1.10 Hybrid π model incorporating Miller capacitor

It is now worth calculating the voltage gain for this new model into a load R_L to observe the break point as this capacitance degrades the frequency response. This will then be converted to S parameters using techniques discussed in Chapter 2 on two port parameters. The voltage across $r_{b'e}$, V_1, in terms of V_{in} is therefore:

$$V_1 = \left[\frac{\dfrac{r_{b'e}}{1 + r_{b'e} C_T j\omega}}{\dfrac{r_{b'e}}{1 + r_{b'e} C_T j\omega} + r_{bb'} + R_s} \right] V_{in} \qquad (1.43)$$

$$= \left[\frac{r_{b'e}}{r_{b'e} + (r_{bb'} + R_s)(1 + r_{b'e} C_T j\omega)} \right] V_{in} \qquad (1.44)$$

Expanding the denominator:

$$V_1 = \left[\frac{r_{b'e}}{r_{b'e} + r_{bb'} + R_s + (r_{bb'} + R_s)(r_{b'e} C_T j\omega)} \right] V_{in} \qquad (1.45)$$

As:

$$\frac{a}{b+c} = \frac{a}{b\left(1 + \dfrac{c}{b}\right)} \qquad (1.46)$$

$$V_1 = \left[\frac{r_{b'e}}{r_{b'e} + r_{bb'} + R_s} \right] \left[\frac{1}{1 + j\omega C_T \dfrac{(r_{bb'} + R_s) r_{b'e}}{r_{b'e} + r_{bb'} + R_s}} \right] V_{in} \qquad (1.47)$$

As:

$$V_{out} = - g_m R_L V_1 \qquad (1.48)$$

the voltage gain is therefore:

$$\frac{V_{out}}{V_{in}} = -\left(g_m R_L\right)\left[\frac{r_{b'e}}{r_{b'e} + r_{bb'} + R_s}\right]\left[\frac{1}{1 + j\omega C_T \dfrac{\left(r_{bb'} + R_s\right)r_{b'e}}{r_{b'e} + r_{bb'} + R_s}}\right] \quad (1.49)$$

Note that:

$$\left[\frac{1}{1 + j\omega C_T \dfrac{\left(r_{bb'} + R_s\right)r_{b'e}}{r_{b'e} + r_{bb'} + R_s}}\right] = \frac{1}{1 + j\omega C_T R} \quad (1.50)$$

where:

$$R = \frac{\left(r_{bb'} + R_s\right)r_{b'e}}{r_{b'e} + r_{bb'} + R_s} \quad (1.51)$$

This is effectively $r_{bb'}$ in series with R_s all in parallel with $r_{b'e}$ which is the effective Thévenin equivalent, total source resistance seen by the capacitor. The first two brackets of equation (1.49) show the DC voltage gain and the third bracket describes the roll-off where:

$$C_T = \left(1 + g_m R_L\right)C_{b'c} + C_{b'e} \quad (1.52)$$

The numerator of the third bracket produces the 3dB point when the imaginary part is equal to one.

$$2\pi f_{3dB}\left[C_T\right]\frac{\left(r_{bb'} + R_s\right)r_{b'e}}{r_{b'e} + r_{bb'} + R_s} = 1 \quad (1.53)$$

The full equation is:

$$2\pi f_{3dB}\left[\left(1 + g_m R_L\right)C_{b'c} + C_{b'e}\right]\frac{\left(r_{bb'} + R_s\right)r_{b'e}}{r_{b'e} + r_{bb'} + R_s} = 1 \quad (1.54)$$

Therefore from equation (1.53):

$$f_{3dB} = \frac{1}{2\pi[C_T]\dfrac{(r_{bb'} + R_s)r_{b'e}}{r_{b'e} + r_{bb'} + R_s}} \tag{1.55}$$

$$f_{3dB} = \frac{r_{b'e} + r_{bb'} + R_s}{2\pi[C_T](r_{bb'} + R_s)r_{b'e}} \tag{1.56}$$

As:

$$C_T = [(1 + g_m R_L)C_{b'c} + C_{b'e}] \tag{1.52}$$

$$f_{3dB} = \frac{r_{b'e} + r_{bb'} + R_s}{2\pi[(1 + g_m R_L)C_{b'c} + C_{b'e}](r_{bb'} + R_s)r_{b'e}} \tag{1.57}$$

Note that using equation (1.51) for R:

$$f_{3dB} = \frac{1}{2\pi[(1 + g_m R_L)C_{b'c} + C_{b'e}]R} \tag{1.58}$$

1.4   *S* Parameter Equations

This equation describing the voltage gain can now be converted to 50Ω S parameters by making $R_s = R_L = 50\Omega$ and calculating S_{21} as the value of V_{out} when V_{in} is set to 2V. The equation and explanation for this are given in Chapter 2. Equation (1.49) therefore becomes:

$$S_{21} = 2(g_m . 50)\left[\frac{r_{b'e}}{r_{b'e} + r_{bb'} + 50}\right]\left[\frac{1}{1 + j\omega C_T \dfrac{(r_{bb'} + 50)r_{b'e}}{r_{b'e} + r_{bb'} + 50}}\right] \tag{1.59}$$

1.5   Example Calculations of S_{21}

It is now worth inserting some typical values, similar to those used when h_{21} was investigated, to obtain S_{21}. Further it will be interesting to note the added effect caused by the feedback capacitor. Take two typical examples of modern RF transistors both with an f_T of 5GHz where one transistor is designed to operate at 10mA and the other at 1mA. The calculations will then be compared with theory in graphical form.

1.5.1   Medium Current RF Transistor – 10mA

Assume that f_T = 5GHz, h_{fe} = 100, I_c=10mA, and the feedback capacitor, $C_{b'c} \approx$ 2pF. Therefore $r_e = 2.5\Omega$ and $r_{b'e} \approx 250\Omega$. Let $r_{bb'} \approx 10\Omega$ for a typical 10mA device. The 3dB point for h_{21} (short circuit current gain) when placed directly in a common emitter circuit is $f_\beta = f_T/h_{fe}$. Therefore f_β = 50MHz.

As:

$$CR = \frac{1}{2\pi f_\beta} \tag{1.24}$$

then $CR = 3.18 \times 10^{-9}$

As h_{21} is the short circuit current gain $C_{b'c}$ and $C_{b'e}$ are effectively in parallel. Therefore:

$$C = C_{b'e} + C_{b'c} = 12.6\text{pF} \tag{1.61}$$

As

$$C_{b'c} \approx 1pF \quad C_{b'e} = 12.6 - 1 = 11.6\text{pF} \tag{1.62}$$

However, C_T for the measurement of S parameters includes the Miller effect because the load impedance is not zero. Thus:

$$C_T = (1 + g_m R_L)C_{b'c} + C_{b'e} = \left(1 + \frac{R_L}{r_e}\right)C_{b'c} + C_{b'e} \tag{1.63}$$

$$C_T = (1+20)\!1 + 11.6 = 32.6\text{pF} \qquad (1.64)$$

As:

$$f_{3dB} = \frac{r_{b'e} + r_{bb'} + R_s}{2\pi[C_T](r_{bb'} + R_s)r_{b'e}} \qquad (1.65)$$

then:

$$f_{3dB} = \frac{250+10+50}{2\pi[32.6\times10^{-12}](10+50)250} = 101\text{MHz} \qquad (1.66)$$

The value of S_{21} at low frequencies is therefore:

$$S_{21} = 2(g_m R_L)\left[\frac{r_{b'e}}{r_{b'e} + r_{bb'} + 50}\right] \qquad (1.67)$$

$$S_{21} = 2(20)\left[\frac{250}{250+10+50}\right] = 32.25 \qquad (1.68)$$

A further modification which will give a more accurate answer is to modify r_e to include the dynamic diode resistance as before but now to include an internal emitter fixed resistance of around 1Ω which is fairly typical for this type of device. This would then make $r_e = 3.5\Omega$ and $r_{b'e} = 350\Omega$. For the same f_T and the same feedback capacitance the calculations can be modified to obtain:

$$C = C_{b'e} + C_{b'c} = 9\text{pF} \quad \text{As } C_{b'c} \approx 1pF \text{ and } C_{b'e} = 9-1 = 8\text{pF} \qquad (1.69)$$

$$C_T = \left(1 + \frac{R_L}{r_e}\right)C_{b'c} + C_{b'e} = (1+14.3)\!1 + 8 = 23.3\text{pF} \qquad (1.70)$$

As:

$$f_{3dB} = \frac{r_{b'e} + r_{bb'} + R_s}{2\pi[C_T](r_{bb'} + R_s)r_{b'e}} \qquad (1.65)$$

then:

$$f_{3dB} = \frac{350 + 10 + 50}{2\pi [23.3\,E - 12](10 + 50)350} = 133\text{MHz} \qquad (1.71)$$

The low frequency S_{21} is therefore:

$$S_{21} = 2.(g_m.50)\left[\frac{r_{b'e}}{r_{b'e} + r_{bb'} + 50}\right] \qquad (1.72)$$

$$S_{21} = 2.(14.3)\left[\frac{350}{350 + 10 + 50}\right] = 24.4 \qquad (1.73)$$

This produces an even more accurate answer. These values are fairly typical for a transistor of this kind, e.g. the BFR92A. This is illustrated in Figure 1.11 where the dotted line is the calculation and the discrete points are measured S parameter data.

Note that the value for the base spreading resistance and the emitter contact resistance can be obtained from the SPICE model for the device where the base series resistance is RB and the emitter series resistance is RE.

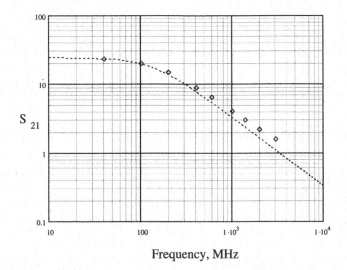

Figure 1.11 S_{21} for a typical bipolar transistor operating at 10mA

1.5.2   Lower Current Device - 1mA

If we now take a lower current device with the parameters $f_T = 5\text{GHz}$, $h_{fe} = 100$, $I_c = 1\text{mA}$, and the feedback capacitor, $C_{b'c} \approx 0.2\text{pF}$, $r_e = 25\Omega$ and $r_{b'e} \approx 2500\Omega$.

Let $r_{bb'} \approx 100\Omega$ for a typical 1mA device. The 3dB point for h_{21} (short circuit current gain) when placed directly in a common emitter circuit is:

$$f_\beta = \frac{f_T}{hfe} \tag{1.22}$$

Therefore $f_\beta = 50\text{MHz}$

$$f_T = h_{fe} \cdot f_\beta = \frac{h_{fe}}{2\pi CR} \tag{1.21}$$

Note also that:

$$CR = \frac{1}{2\pi f_\beta} \tag{1.74}$$

Therefore $CR = 3.18 \times 10^{-9}$

As h_{21} is the short circuit current gain, $C_{b'c}$ and $C_{b'e}$ are effectively in parallel. Therefore:

$$C = C_{b'e} + C_{b'c} = 1.26\text{pF} \tag{1.75}$$

As $C_{b'c} \approx 0.2\text{pF}$

$$C_{b'e} = 1.26 - 0.2 = 1.06\text{pF} \tag{1.76}$$

However, C_T for the measurement of S parameters includes the Miller effect because the load impedance is not zero. Therefore:

$$C_T = (1 + g_m R_L)C_{b'c} + C_{b'e} = \left(1 + \frac{R_L}{r_e}\right)C_{b'c} + C_{b'e} \tag{1.77}$$

$$C_T = (1+2)0.2 + 1.06 = 1.66\text{pF} \tag{1.78}$$

As:

$$f_{3dB} = \frac{r_{b'e} + r_{bb'} + R_s}{2\pi[C_T](r_{bb'} + R_s)r_{b'e}} \tag{1.65}$$

then:

$$f_{3dB} = \frac{2500 + 100 + 50}{2\pi[1.66\times10^{-12}](100 + 50)2500} = 678\text{MHz} \tag{1.79}$$

The low frequency value for S_{21} is therefore:

$$S_{21} = 2(g_m.50)\left[\frac{r_{b'e}}{r_{b'e} + r_{bb'} + 50}\right] \tag{1.72}$$

$$S_{21} = 2(2)\left[\frac{2500}{2500 + 100 + 50}\right] = 3.77 \tag{1.80}$$

As before an internal emitter fixed resistance can be incorporated. For these lower current smaller devices a figure of around 8Ω is fairly typical. This would then make $r_e = 33\Omega$ and $r_{b'e} = 3300\Omega$. For the same f_T and the same feedback capacitance the calculations can be modified to obtain:

$$C = C_{b'e} + C_{b'c} = 0.96\text{pF} \tag{1.81}$$

As: $C_{b'c} \approx 0.2\text{pF}$

$$C_{b'e} = 0.96 - 0.2 = 0.76\text{pF} \tag{1.82}$$

$$C_T = \left(1 + \frac{R_L}{r_e}\right)C_{b'c} + C_{b'e} = (1 + 1.52)0.2 + 0.76 = 1.26\text{pF} \tag{1.83}$$

As:

$$f_{3dB} = \frac{r_{b'e} + r_{bb'} + R_s}{2\pi[C_T](r_{bb'} + R_s)r_{b'e}}$$ (1.65)

then:

$$f_{3dB} = \frac{3300 + 100 + 50}{2\pi[1.26 \times 10^{-12}](100+50)3300} = 880\text{MHz}$$

The low frequency S_{21} is therefore:

$$S_{21} = 2(g_m.50)\left[\frac{r_{b'e}}{r_{b'e} + r_{bb'} + 50}\right]$$ (1.84)

$$S_{21} = 2(1.52)\left[\frac{3300}{3300 + 100 + 50}\right] = 2.91$$ (1.85)

These values are fairly typical for a transistor of this kind operating at 1mA. Calculated values for S_{21} and measured data points for a typical low current device, such as the BFG25A, are shown in Figure 1.12.

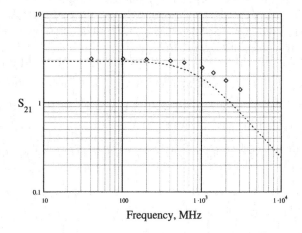

Figure 1.12 S_{21} for a typical bipolar transistor operating at 1mA

It has therefore been shown that by using a simple set of models a significant amount of accurate information can be gained by using f_T, h_{fe}, the feedback capacitance and the operating current.

In summary the low frequency value of S_{21} is therefore:

$$S_{21} = 2(g_m.R_L)\left[\frac{r_{b'e}}{r_{b'e} + r_{bb'} + 50}\right] \approx 2g_m R_L \approx 2\frac{R_L}{r_e} \qquad (1.86)$$

and the 3dB point is:

$$f_{3dB} = \frac{r_{b'e} + r_{bb'} + R_s}{2\pi[C_T](r_{bb'} + R_s)r_{b'e}} \qquad (1.65)$$

where:

$$C_T = \left(1 + \frac{R_L}{r_e}\right)C_{b'c} + C_{b'e} \qquad 1.83)$$

1.6   Common Base Amplifier

It is now worth investigating the common base amplifier where the base is grounded, the input is connected to the emitter and the output is connected to the collector. This is most easily shown using the T model as shown in Figure 1.13. The π model is included in Figure 1.14.

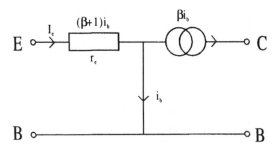

Figure 1.13 T model for common base

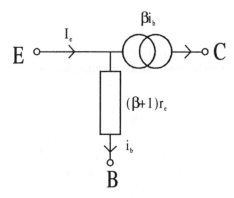

Figure 1.14 The π model for common base

The input impedance at the emitter is:

$$Z_{in} = r_e \qquad (1.87)$$

The voltage gain is therefore:

$$\frac{V_{out}}{V_{in}} = \frac{\beta i_b R_L}{(\beta + 1) i_b r_e} \approx \frac{R_L}{r_e} \qquad (1.88)$$

Note that this is non-inverting. The current gain is:

$$\frac{I_{out}}{I_{in}} = \frac{\beta i_b}{(\beta + 1) i_b} = \alpha \approx 1 \qquad (1.89)$$

The major features of this type of circuit are:

1.   The negligible feedback C. Further the grounded base also partially acts as a screen for any parasitic feedback.

2.   No current gain, so the amplifier gain can only be obtained through an increase in impedance from the input to the output.

3.   Low input resistance.

4.   High output impedance.

1.7   Cascode

It is often useful to combine the features of the common emitter and common base modes of operation in a Cascode transistor configuration. This is shown in Figure 1.15.

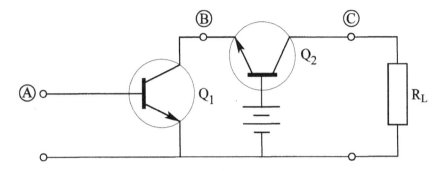

Figure 1.15 Cascode configuration

There are a number of useful features about the Cascode:

1.   The capacitance between B and C is very low. Further as the base of Q_2 is grounded this acts as a screen to any parasitic components.

2.   The input impedance Z_{in} at B of Q_2 is:

$$r_e = \frac{1}{g_m} = \frac{25}{I_e (\text{mA})} \qquad (1.90)$$

This is quite low.

3.   This means that the Miller effect on Q_1 is small. Further the voltage gain of Q_1 is:

$$\frac{V_B}{V_A} = -g_m r_e = -1 \qquad (1.91)$$

as the current flowing in both transistors is the same. In fact this puts two times $C_{b'c}$ across the input of Q_1 when the Miller effect is included.

4.   The input impedance at A of Q_1 is:

$$Z_{in} = \frac{V_{in}}{i_b} = (1 + \beta)r_e = (1 + \beta)\frac{25}{I_{mA}}$$ (1.92)

5.   The voltage gain is:

$$\frac{V_{out}}{V_{in}} = -g_m R_L = -\frac{R_L}{r_e}$$ (1.93)

6.   The current gain is β.

This device therefore offers the input impedance of a single transistor common emitter stage with an input capacitance of:

$$C_T = \left[(1 + g_m R_L)C_{b'c} + C_{b'e}\right] = \left[(2)C_{b'c} + C_{b'e}\right]$$ (1.94)

but with very low feedback between the input and output typically of the order of 20fF. A diagram of the typical biasing for a Cascode is shown in Figure 1.16. The important features here are the decoupling on the base of Q_2 to ensure a good AC ground and that the collector base voltage of Q_1 should be sufficiently large to ensure low feedback capacitance.

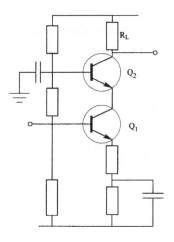

Figure 1.16 Cascode bias circuit

1.8   Large Signal Modelling – Harmonic and Third Order Intermodulation Distortion

So far linear models have been presented and it has been shown how significant information can be obtained from simple algebraic analysis. Here the simple model, shown earlier in Figure 1.1b, will now be used to demonstrate large signal modelling of a bipolar transistor in both common emitter and differential mode. The harmonic and third order intermodulation products will be deduced. It will then be shown how differential circuits suppress even order terms such as the second and fourth harmonics.

The simple model consists of a current controlled current source in the base collector region and a diode in the base emitter region. The distortion in the emitter current will be analysed. The collector current is then assumed to be αI_E where $\alpha = \beta/(\beta + 1)$. It will be assumed that the collector load circuit will not cause significant other products. This is a reasonable assumption for small voltage swings well away from collector saturation.

1.8.1   Common Emitter Distortion

Taking the transistor in common emitter mode and applying a DC bias and AC signal, the emitter current becomes:

$$I_E = I_{ES}\left(\exp\left(\frac{V_{BIAS} + V\sin\omega t}{V_T} \right) - 1 \right)$$

(1.95)

where $V_T = kT/e \approx 25\text{mV}$ at room temperature. Separating the DC bias and the AC signal components into separate terms [2]:

$$I_E = I_{ES} \exp\left(\frac{V_{BIAS}}{V_T} \right) \cdot \exp\left(\frac{V\sin\omega t}{V_T} \right) - I_{ES} \approx I_{DC0} \exp\left(\frac{V\sin\omega t}{V_T} \right)$$

(1.96)

where I_{DC0} is the quiescent current when there is no (or small) AC signal. If we expand the exponential term for the AC signals using the first four terms of the series expansion then:

$$I_E = I_{DC0}\left(1 + \left(\frac{V}{V_T}\right)\sin \omega t + \frac{1}{2}\left(\frac{V}{V_T}\right)^2 \sin^2 \omega t + \right.$$

(1.97)

$$\left. + \frac{1}{6}\left(\frac{V}{V_T}\right)^3 \sin^3 \omega t + \frac{1}{24}\left(\frac{V}{V_T}\right)^4 \sin^4 \omega t\right)$$

Expanding the higher order terms to obtain the harmonic frequencies gives:

$$I_E = I_{DC0}\left(1 + \left(\frac{V}{V_T}\right)\sin \omega t + \frac{1}{2}\left(\frac{V}{V_T}\right)^2\left(\frac{1 - \cos 2\omega t}{2}\right) + \right.$$

(1.98)

$$+ \frac{1}{6}\left(\frac{V}{V_T}\right)^3 \frac{3\sin \omega t - \sin 3\omega t}{4} +$$

$$\left. + \frac{1}{24}\left(\frac{V}{V_T}\right)^4 \frac{1.5 - 2\cos 2\omega t - 0.5\cos 4\omega t}{4}\right)$$

(1.98)

There are a number of features that can be seen directly from this equation:

1.   All the harmonics exist where, for example, the second harmonic is produced from the square law term.

2.   The even order terms produce even order harmonics and terms at DC so the DC bias increases with the signal amplitude.

3.   The odd order terms produce signals at the fundamental frequency as well as at the harmonic frequency so even if the harmonics are filtered out there is also non-linearity in the fundamental term.

4.   The distortion products are independent of quiescent current for small signal levels assuming that the emitter to collector transfer function is linear and $V < V_{BIAS}$.

5.   The DC term is therefore dependent on I_{DC0}, the second and fourth order
         terms:

$$I_E = I_{DC0}\left(1 + \frac{1}{4}\left(\frac{V}{V_T}\right)^2 + \frac{1}{64}\left(\frac{V}{V_T}\right)^4\right)$$ (1.99)

Note that this analysis is still an approximation because transistors are usually
biased via resistors, which means that the assumptions produce errors as the signals
increase.

The signal distortion is usually quoted in terms of a ratio of the unwanted
signal over the original input signal and therefore this equation should be
normalised by dividing by V/V_T. The order of the amplitude term is therefore one
order below the order of the distortion. Therefore the distortion products
normalised to the fundamental frequency become:

Fundamental   1   + 3rd order term $\frac{1}{8}\left(\frac{V}{V_T}\right)^2$ (1.100)

Second Harmonic $\frac{1}{4}\frac{V}{V_T}\cos 2\omega t$ + 4th order term $\frac{1}{48}\left(\frac{V}{V_T}\right)^3$ (1.101)

Third Harmonic $\frac{1}{24}\left(\frac{V}{V_T}\right)^2 \sin 3\omega t$ (1.102)

Fourth Harmonic $\frac{1}{192}\left(\frac{V}{V_T}\right)^3 \cos 4\omega t$ (1.103)

1.8.2   Third Order Intermodulation Products

Another common method for testing distortion in circuits is to apply two tones
closely spaced apart and then to investigate terms which are produced close to
these tones. By making these tones equal in amplitude the beat signal swings from
0 to ± 2 times the voltage of each signal. This can therefore be used to test the
circuit over its full dynamic range. It is known that distortion caused by the third
order term produces in-band signals so let us investigate this by analysis.

Let the input signal be two sine waves of equal amplitude:

$$A = \frac{e}{V_T} \sin \omega_1 t \ \ and \ \ B = \frac{e}{V_T} \sin \omega_2 t \tag{1.104}$$

As:

$$(A+B)^3 = A^3 + B^3 + 3A^2B + 3AB^2 \tag{1.105}$$

$$\left(\frac{e}{V_T} \sin \omega_1 t + \frac{e}{V_T} \sin \omega_2 t \right)^3 = \tag{1.106}$$

$$\left(\frac{e}{V_T} \right)^3 \left(\sin^3 \omega_1 t + \sin^3 \omega_2 t + 3 \sin^2 \omega_1 t \sin \omega_2 t + 3 \sin \omega_1 t \sin^2 \omega_2 t \right) \tag{1.106}$$

Expanding these terms produces (1.107):

$$\frac{e}{V_T}^3 \left(\sin^3 \omega_1 t + \sin^3 \omega_2 t + \frac{3}{2} [(1 - \cos 2\omega_1 t) \sin \omega_2 t + (1 - \cos 2\omega_2 t) \sin \omega_1 t] \right)$$

As:

$$-\cos 2\omega_1 t \sin \omega_2 t = \frac{1}{2} [\sin(2\omega_1 - \omega_2)t - \sin(2\omega_1 + \omega_2)t] \tag{1.108}$$

and:

$$-\cos 2\omega_2 t \sin \omega_1 t = \frac{1}{2} [\sin(2\omega_2 - \omega_1)t - \sin(2\omega_2 + \omega_1)t] \tag{1.109}$$

the products caused by the third order terms are therefore:

$$
\left(\frac{e}{V_T}\right)^3
\left(
\begin{array}{l}
\dfrac{3}{2}\left(\sin\omega_1 t + \sin\omega_2 t\right) + \sin^3\omega_1 t + \sin^3\omega_2 t \\[2mm]
+ \dfrac{3}{4}\left[\sin(2\omega_1 - \omega_2)t + \sin(2\omega_2 - \omega_1)t\right] \\[2mm]
- \dfrac{3}{4}\left[\sin(2\omega_1 + \omega_2)t + \sin(2\omega_2 + \omega_1)t\right]
\end{array}
\right)
\qquad (1.110)
$$

Taking just the inband terms in the middle row of equation (1.110), the amplitude of the terms will therefore be:

$$
\left(\frac{1}{6}\cdot\frac{3}{4}\right) \qquad\qquad (1.111)
$$

Therefore:

$$
\frac{1}{8}\cdot\left(\frac{e}{V_T}\right)^3 Sin(2\omega_1 - \omega_2)t + Sin(2\omega_2 - \omega_1)t \qquad (1.112)
$$

Normalising this to the fundamental input tones by dividing by e/V_T produces:

$$
\frac{1}{8}\cdot\left(\frac{e}{V_T}\right)^2 \sin(2\omega_1 - \omega_2)t + \sin(2\omega_2 - \omega_1)t \qquad (1.113)
$$

Two tones each at 1mV would therefore produce a distortion level of 0.02% and two tones at 10mV would produce 2% TIR.

Therefore from this equation a number of parameters can be deduced

1.   Third order intermodulation distortion applied to two tones produces in-band signals which are at frequencies of $(2\omega_1 - \omega_2)$ and $(2\omega_2 - \omega_1)$. For example, two tones at 100 and 101MHz will produce distortion products at 99 and 102MHz. Note similarly that fifth order intermodulation products produce $(3\omega_1 - 2\omega_2)$ and $(3\omega_2 - 2\omega_1)$.

2.   The amplitude of the third order terms changes with a cubic law which is three times the rate in dB. For example, a 1dB change in input signal level will cause 3dB change in distortion product level. Similar rules apply to

fifth and seventh order terms which can also come in band. A 1 dB change in input level will cause a 5dB and a 7dB change in distortion level respectively. Note that this can be used as a means to determine the order of intermodulation distortion during faultfinding. The author has often found higher order distortion products in multi-loop frequency synthesisers and spectrum analysers.

3.   The change between the fundamental and third order terms changes by a square law such that a 1dB change in input signal produces a 2dB change in the difference between the fundamental signal level and the distortion products.

Values of harmonic and third order intermodulation percentage distortion for a variety of input voltages are given in Table 1.1. For the third order intermodulation distortion, the input voltage for each tone is the same as the single tone value for the harmonic distortion. In other words, the third order distortion for two tones each of value 1mV is 0.02%.

Table 1.1 Harmonic and third order intermodulation distortion for a bipolar transistor

ωt	0.1mV	0.5mV	1mV	5mV	10mV
TIR	$2 \times 10^{-6}\%$	$5 \times 10^{-5}\%$	0.02%	0.5%	2%
$2\omega t$	0.1%	0.5%	1%	5%	10%
$3\omega t$	$6.7 \times 10^{-5}\%$	$1.7 \times 10^{-3}\%$	$6.7 \times 10^{-3}\%$	0.17%	0.67%
$4\omega t$	$3.3 \times 10^{-8}\%$	$4.2 \times 10^{-5}\%$	$3.3 \times 10^{-5}\%$	$4.2 \times 10^{-3}\%$	0.033%

1.8.3   Differential Amplifier

It is now worth investigating the same model when connected in differential mode as shown in Figure 1.17.

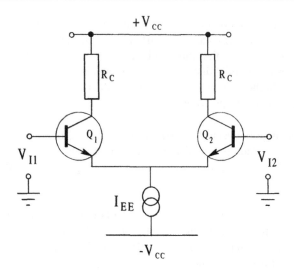

Figure 1.17 Differential amplifier

Taking each transistor in turn [2]:

$$I_{E1} = I_{ES1} \exp\left(\frac{V_{BE1}}{V_T}\right)$$ (1.114)

$$I_{E2} = I_{ES2} \exp\left(\frac{V_{BE2}}{V_T}\right)$$ (1.115)

Assume that the transistors are matched such that $I_{ES1} = I_{ES2}$ then:

$$\frac{I_{E1}}{I_{E2}} = \exp\left(\frac{V_{BE1} - V_{BE2}}{V_T}\right)$$ (1.116)

As the total voltage around a closed loop is zero then:

$$V_{I1} - V_{BE1} + V_{BE2} - V_{I2} = 0$$ (1.117)

Therefore:

$$V_{BE1} - V_{BE2} = V_{I1} - V_{I2} = V_{ID} \tag{1.118}$$

and:

$$\frac{I_{E1}}{I_{E2}} = \exp\left(\frac{V_{ID}}{V_T}\right) \tag{1.119}$$

Note also that the current through the current source equals the sum of the emitter currents:

$$I_{EE} = I_{E1} + I_{E2} \tag{1.120}$$

$$I_{E1} = I_{E2} \exp\left(\frac{V_{ID}}{V_T}\right) = \left(I_{EE} - I_{E1}\right)\exp\left(\frac{V_{ID}}{V_T}\right) \tag{1.121}$$

$$I_{E1}\left(1 + \exp\left(\frac{V_{ID}}{V_T}\right)\right) = I_{EE}\exp\left(\frac{V_{ID}}{V_T}\right) \tag{1.122}$$

Therefore:

$$I_{E1} = \frac{I_{EE}\exp\left(V_{ID}/V_T\right)}{1 + \exp\left(V_{ID}/V_T\right)} \tag{1.123}$$

Multiplying top and bottom by $\exp\left(-\dfrac{V_{ID}}{V_T}\right)$: $\tag{1.124}$

$$I_{E1} = \frac{I_{EE}}{1 + \exp\left(-\dfrac{V_{ID}}{V_T}\right)} \tag{1.125}$$

Similarly:

$$I_{E2} = \frac{I_{EE}}{1 + \exp\left(\dfrac{V_{ID}}{V_T}\right)}$$

(1.126)

To calculate the differential output voltage:

$$V_{O1} = V_{CC} - I_{C1}R_C$$

(1.127)

$$V_{O2} = V_{CC} - I_{C2}R_C$$

(1.128)

Therefore the differential output voltage is:

$$V_{OD} = V_{O1} - V_{O2} = I_{C2}R_C - I_{C1}R_C$$

(1.129)

$$V_{OD} = \alpha I_{EE} R_C \left(\frac{1}{\exp\left(\dfrac{V_{ID}}{V_T}\right) + 1} - \frac{1}{\exp\left(-\dfrac{V_{ID}}{V_T}\right) + 1} \right)$$

(1.130)

where α converts the emitter current to the collector current so:

$$V_{OD} = \alpha I_{EE} R_C \left(\frac{1}{\exp\left(\dfrac{V_{ID}}{V_T}\right) + 1} - \frac{\exp\left(\dfrac{V_{ID}}{V_T}\right)}{1 + \exp\left(\dfrac{V_{ID}}{V_T}\right)} \right)$$

(1.131)

$$V_{OD} = \alpha I_{EE} R_C \left(\frac{1 - \exp\left(\dfrac{V_{ID}}{V_T}\right)}{1 + \exp\left(\dfrac{V_{ID}}{V_T}\right)} \right)$$

(1.132)

As:

$$\tanh(x) = \frac{\exp(2x) - 1}{\exp(2x) + 1}$$

(1.133)

$$V_{OD} = \alpha I_{EE} R_C \tanh\left(\frac{-V_{ID}}{2V_T}\right)$$

(1.134)

As:

$$\tanh(x) = x - \frac{1}{3}x^3 + \frac{2}{15}x^5 - \frac{17}{315}x^7$$

(1.135)

and assuming that V_{ID} is a sine wave, then:

$$V_{OD} = \alpha I_{EE} R_C \left(\left(\frac{-V_{ID}}{2V_T}\right)\sin \omega t + \frac{1}{3}\left(\frac{V_{ID}}{2V_T}\right)^3 \sin^3 \omega t - ...\right)$$

(1.136)

Taking the ½ outside the bracket and expanding the third order term then:

$$V_{OD} = \frac{\alpha I_{EE} R_C}{2}\left(\left(\frac{-V_{ID}}{V_T}\right)\sin \omega t + \frac{1}{12}\left(\frac{V_{ID}}{V_T}\right)^3 \left(\frac{3\sin \omega t - \sin 3\omega t}{4}\right) - ...\right)$$

(1.137)

As a check for this equation the small signal gain can be calculated from the linear terms as:

$$\frac{V_{OD}}{V_{ID}} = -\frac{\alpha I_{EE} R_C}{2V_T} = -g_m R_C$$

(1.138)

Because the emitter current in each transistor is:

$$\frac{I_{EE}}{2}$$

(1.139)

then:

$$g_m = \frac{I_{EE}}{2V_T} \approx \frac{1}{re} \tag{1.140}$$

Equations (1.135) and (1.137) demonstrate that:

1.   Only odd harmonics exist owing to circuit symmetry.

2.   The DC bias does not change.

3.   These circuits make good symmetrical limiters.

To calculate the harmonic distortion and third order intermodulation distortion it is necessary to normalise the equations as before by dividing by V_{ID}/V_T. The third harmonic distortion ratio is therefore:

$$\frac{1}{48}\left(\frac{V_{ID}}{V_T}\right)^2 \sin 3\omega t \tag{1.141}$$

The third order intermodulation distortion ratio is therefore:

$$\frac{1}{16}\cdot\left(\frac{e}{V_T}\right)^2 \sin(2\omega_1 - \omega_2)t + \sin(2\omega_2 - \omega_1)t \tag{1.142}$$

1.9   Distortion Reduction Using Negative Feedback

Harmonic and third order intermodulation distortion can be reduced by applying negative feedback. In common emitter and differential amplifiers this is most easily achieved by inserting resistors in the emitter circuit. The estimation of the reduction in distortion can be rather difficult to analyse. It is therefore worth performing a simple generic calculation to see the effect of negative feedback on distortion products.

Take a general model of a negative feedback system as shown in Figure 1.18. Assume that the amplifier introduces a small error ε and then examine the effect of negative feedback.

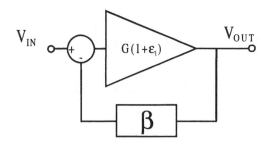

Figure 1.18   Model of distortion reduction with negative feedback

The voltage transfer characteristic can be calculated as:

$$\frac{V_{OUT}}{V_{IN}} = \frac{G(1+\varepsilon)}{1 + \beta G(1+\varepsilon)}$$

(1.143)

$$\frac{V_{OUT}}{V_{IN}} = \frac{G(1+\varepsilon)}{(1 + \beta G) + \beta G \varepsilon}$$

(1.144)

Dividing the top and bottom by $(1 + \beta G)$:

$$\frac{V_{OUT}}{V_{IN}} = \frac{\dfrac{G(1+\varepsilon)}{(1+\beta G)}}{1 + \dfrac{\beta G \varepsilon}{(1+\beta G)}}$$

(1.145)

For:

$$\frac{\beta G \varepsilon}{(1+\beta G)} \ll 1$$

(1.146)

$$\frac{V_{OUT}}{V_{IN}} = \frac{G}{(1+\beta G)}(1+\varepsilon)\left(1 - \frac{\beta G \varepsilon}{(1+\beta G)}\right)$$

(1.147)

$$\frac{V_{OUT}}{V_{IN}} = \frac{G}{(1+\beta G)}\left(1+\varepsilon - \frac{\beta G\varepsilon}{1+\beta G} - \frac{\beta G\varepsilon^2}{1+\beta G}\right)$$

(1.148)

$$\frac{V_{OUT}}{V_{IN}} = \frac{G}{(1+\beta G)}\left(1+\frac{\varepsilon(1+\beta G)}{1+\beta G} - \frac{\beta G\varepsilon}{1+\beta G} - \frac{\beta G\varepsilon^2}{1+\beta G}\right)$$

(1.149)

Ignoring square law terms then:

$$\frac{V_{OUT}}{V_{IN}} = \frac{G}{(1+\beta G)}\left(1+\frac{\varepsilon}{1+\beta G}\right)$$

(1.150)

The error is therefore reduced by:

$$\frac{1}{1+\beta G} = \frac{G}{1+\beta G}\cdot\frac{1}{G}$$

(1.151)

Which is equal to the ratio of:

$$\frac{Closed\,loop\,gain}{Open\,loop\,gain}$$

(1.152)

For example, if it is necessary reduce the distortion in an amplifier with a open loop gain of 20 by a factor of 2 then the closed loop gain after feedback should be reduced to 10.

It is now useful to look at the dual gate MOSFET which is an integrated Cascode of two FETs with the added feature that the bias on gate 2 can be used to vary the gain between gate 1 and the output drain by up to 50dB. Initially it is worth investigating the single gate MOSFET.

1.10   RF MOSFETs

The FET is a device in which the current in the output circuit is controlled purely by the voltage at the input junction. This can therefore be modelled as a voltage controlled current source. The fundamental principle of operation is different from the bipolar transistor, in that the input voltage controls the charge density and hence current flow in the channel between the output terminals. In the case of

MOSFETs, the input or gate terminal is insulated from the channel (usually) by a layer of silicon dioxide. (In the junction FET insulation is produced by a reverse biased junction.) FETs are majority carrier only devices, thus avoiding problems associated with minority carrier storage which can limit high frequency performance. However, the g_m for these devices is lower than that for bipolar devices.

The dual gate MOSFET is often used for amplifier and mixer applications at VHF and UHF and therefore this will be described after a description of the single gate MOSFET.

1.10.1   Small Signal Analysis

The small signal analysis of the MOSFET follows the basic procedure as outlined for the bipolar transistor in Section 1.2. The transistor is generally operated in 'common source' mode, and a simple small signal equivalent circuit is shown in Figure 1.19.

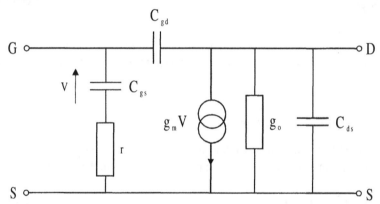

Figure 1.19 Small signal equivalent circuit of a MOSFET

The input consists of a capacitor with a small loss resistance which can usually be ignored for RF applications. The output current is controlled by the voltage across the capacitor C_{gs} with a forward transconductance g_m.

1.10.2   Capacitive Terms

The physical capacitance of the gate contact constitutes the input gate-source capacitance, C_{gs}.

$$C = \frac{\varepsilon \times \text{gate area}}{\text{oxide thickness}} \qquad\qquad (1.153)$$

The feedback capacitance, C_{gd} is usually smaller than the input capacitance. The Miller effect can multiply this significantly in voltage amplifiers, and may cause instability. The output capacitance can be of a similar magnitude to the gate capacitance in some transistors. The output resistance can be found from the slope of the output I_D versus V_{DS} characteristic at the bias point.

1.10.3   Transition Frequency f_T

The transition frequency, f_T, can be determined in a similar way to bipolar transistors by calculating the unity current gain transition frequency:

$$f_T = \frac{g_m}{2\pi C_{//}} \quad \text{(Hz)} \qquad\qquad (1.154)$$

where the capacitance $C_{//}$ is the parallel combination of $C_{gs} + C_{gd}$. Note C_{gd} is usually less than C_{gs} for MOSFETs. In fact for **dual gate** MOSFETs (discussed next) it can be 100 times less.

1.10.4   MOSFET y-parameters

The MOSFET y parameters (short circuit parameters) are given below for this equivalent circuit as these are often the parameters given in the data sheet:

$$y_{11} = j\omega\left(C_{gs} + C_{gd}\right) \qquad\qquad (1.155)$$

$$y_{21} = \frac{g_m}{1 + j\omega C_{gs}r} - j\omega C_{gd} \qquad\qquad (1.157)$$

$$y_{12} = -j\omega C_{gd} \qquad\qquad (1.158)$$

$$y_{22} = g_O + j\omega\left(C_{ds} + C_{gd}\right) \qquad\qquad (1.159)$$

It will be shown in Chapters 2 and 6 how parameter manipulation can be used to obtain the large signal model for a power RF FET.

1.10.5   Dual Gate MOSFETs

A schematic diagram of a dual gate MOSFET is shown in Figure 1.20. The two gates are placed side by side above the channel of the transistor. The electrons must therefore pass under both gates while travelling from the source to the drain. The device behaves like a cascode connected pair of MOSFETs. As expected in a cascode stage, the reverse transfer capacitance is extremely small, around 20fF, improving the stability of amplifier circuits, although the high input and output impedance tends to reduce this advantage. Diodes are incorporated to offer gate breakdown protection.

A further feature is that the bias on gate 2 can be used to vary the gain between gate 1 and the output drain by up to 50dB. This has immediate application in gain control for amplifiers requiring AGC. It can also be used directly to create multiplication and hence mixing by driving gate two with a high power LO and applying the RF signal to gate one. The transadmittance y_{21} for a typical dual gate MOSFET is shown as a function of the two gate voltages in Figure 1.21. From this characteristic the optimum operating points for amplifier and mixer application can be found.

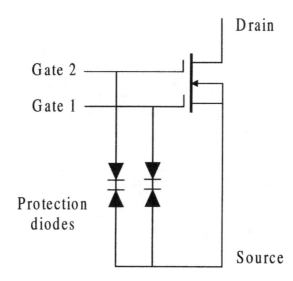

Figure 1.20 Dual gate MOSFET with gate protection diodes

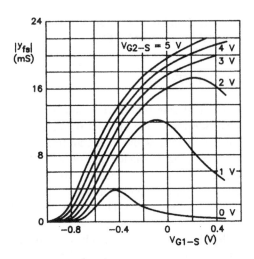

Figure 1.21 Typical transadmittance y_{21} as a function of the two gate biases. (Reproduced with permission from Philips, using data book SC07 on Small Signal Field Effect Transistors.)

A typical circuit for a dual gate MOSFET amplifier is shown in Figure 1.22. This shows the DC biasing network and tapped C matching networks on the input and output. The matching networks will be described in Chapter 3 on amplifier design.

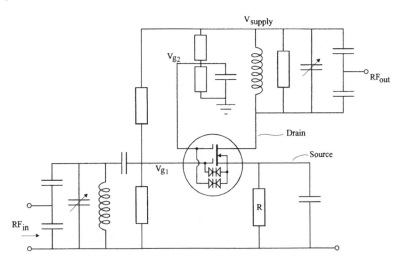

Figure 1.22 Circuit for a dual gate MOSFET amplifier

To calculate the bias current for the device the I_D vs V_{GIS} characteristics are used as shown in Figure 1.23. It can be seen that a V_{GIS} bias of zero volts produces a drain current of 10mA for this device. Note however the large variation from device to device. This is a typical problem in MOSFET devices. This change can be reduced by having a large voltage drop across the source resistor to operate more like a current source. This can be quantified using load lines. From Figure 1.21 it can be seen that V_{G2S} should be set to at least 4 volts for high gain between gate 1 and the output. A typical I_D vs V_{DS} characteristic is shown in Figure 1.24 where the output current vs V_{GIS} is also given. Note that V_{G2S} is kept constant here at 4 volts.

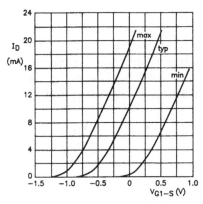

Figure 1.23 Typical I_D vs V_{GIS} for $V_{G2S} = 4$V. (Reproduced with permission from Philips, using data book SC07 on Small Signal Field Effect Transistors.)

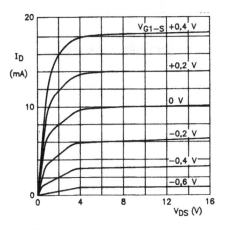

Figure 1.24 Typical I_D vs V_{DS} for varying V_{GIS}. $V_{G2S} = 4$V. (Reproduced with permission from Philips, using data book SC07 on Small Signal Field Effect Transistors.)

1.11   Diode Detectors

It is often necessary to detect the RF power in a system. This is usually achieved using a diode detector if a moderate sensitivity is required. These detectors are most commonly used in one of two modes:

1.   The square law mode $V_{out} \propto P_{in}$ for low power detection from around -20 to -50 dBm.

2.   The linear mode for AM demodulation where $V_{out} \propto V_{in}$ for power levels above around –15dBm.

For the detection of lower power levels, coherent detection is often used as in a spectrum analyser or a superheterodyne receiver.

This section will describe the theory and operation of a typical diode detector. The basic structure for a diode detector is shown in Figure 1.25. This consists of a DC return (the inductor), a diode and capacitive load.

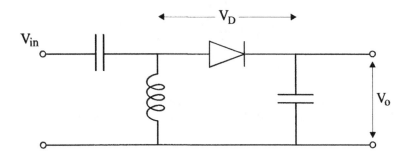

Figure 1.25   Diode detector

Recalling the diode equation for current vs voltage bias:

$$I = I_0\left(\exp\left(\frac{eV}{kT} \right) - 1 \right)$$
                                                                                                (1.159)

As: $1/V_T = e/kT$, then:

$$I = I_0\left(\exp\left(\frac{V}{V_T} \right) - 1 \right)$$
                                                                                                (1.160)

It is worth expanding this exponential equation as a Taylor Series, therefore:

$$I = I_0 \left(1 + \frac{1}{1!}\left(\frac{V}{V_T}\right) + \frac{1}{2!}\left(\frac{V}{V_T}\right)^2 + \frac{1}{3!}\left(\frac{V}{V_T}\right)^3 + \frac{1}{4!}\left(\frac{V}{V_T}\right)^4 + .. -1 \right) \quad (1.161)$$

The ones cancel. If we ignore the terms above second order a very close approximation to the diode response can be obtained. The voltage across the diode with an applied RF voltage after steady state has occurred is (from Figure 1.25):

$$V_D = V_{in}\cos\omega t - V_o \quad\quad\quad (1.162)$$

The first order term is then:

$$\frac{V_D}{V_T} = \frac{V_{in}\cos\omega t - V_o}{V_T} \quad\quad\quad (1.163)$$

The second order terms is then:

$$\frac{1}{2}\left(\frac{V_D}{V_T}\right)^2 = \frac{1}{2V_T^{\,2}}(V_{in}\cos\omega t - V_0)(V_{in}\cos\omega t - V_0)$$

$$\quad\quad\quad (1.164)$$

$$= \frac{1}{2V_T^{\,2}}\left(V_0^{\,2} - 2V_0 V_{in}\cos\omega t + V_{in}^{\,2}\cos^2\omega t\right)$$

Expanding the square law term:

$$\frac{1}{2}\left(\frac{V_D}{V_T}\right)^2 = \frac{1}{2V_T^{\,2}}\left(V_0^{\,2} - 2V_0 A\cos\omega t + \frac{V_{in}^{\,2}}{2}(1 + \cos 2\omega t)\right) \quad (1.165)$$

Ignoring the AC terms and assuming that under a steady state RF input the current becomes equal to zero then:

$$I = I_0 \left(-\frac{V_0}{V_T} + \frac{V_0^2}{2V_T} + \frac{V_{in}^2}{4V_T^2} \right) = 0 \qquad (1.166)$$

At room temperature:

$$\frac{1}{V_T} = \frac{e}{kT} = 40 \qquad (1.167)$$

$$I_0 \left(-40V_0 + 800V_0^2 + 400V_{in}^2 \right) = 0 \qquad (1.168)$$

This simplifies to:

$$\left(-V_0 + 20V_0^2 + 10V_{in}^2 \right) = 0 \qquad (1.169)$$

At low power levels this simplifies to:

$$V_0 = 10V_{in}^2 \qquad (1.170)$$

A plot of V_o vs V_{in} (including the high order effects) is shown in Figure 1.26. A plot of V_0 against the power into a 50Ω load where the power:

$$P = \frac{V_{in}^2}{100} \qquad (1.171)$$

is shown in Figure 1.27. Note that for broadband operation a resistor can be placed across the inductor to provide 50Ω input impedance. It is also possible to use a matching network using a transformer or resonant networks. It can be seen that the detector operates as a square law detector:

$$V_0 \propto V_{in}^2 \qquad (1.172)$$

i.e. linear with power:

$$V_0 \propto P \qquad (1.173)$$

up to peak to peak input voltages around 30mV and input powers around -15dBm. Above this value the curve tends to:

$$V_0 \propto V_{in} \qquad\qquad (1.174)$$

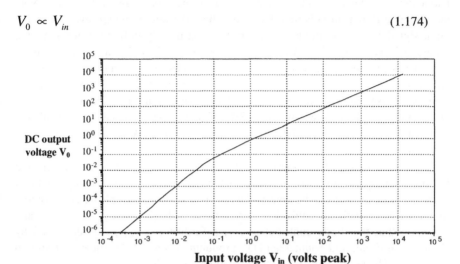

Figure 1.26 V_0 vs V_{in} for a diode detector

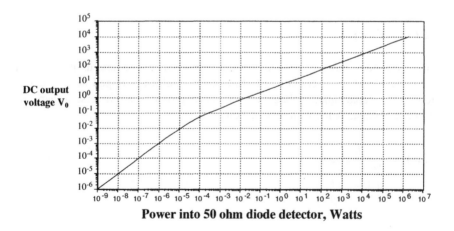

Figure 1.27 V_0 vs P for a diode detector

1.11.1   Minimum Detectable Signal Level - Tangential Sensitivity

Diode detectors have a limited sensitivity due to the square law performance. Note that this is different in coherent detection where the output signal level tracks the input signal level. The diode noise sensitivity is usually defined in terms of the Tangential Sensitivity. This was originally defined as the point at which the signal level was equal to the peak of the noise level as observed on an oscilloscope. Of course this is rather arbitrary due to the statistics of noise. Remember that for Gaussian noise the rms value is one standard deviation and that the noise exceeds three standard deviations around 0.3% of the time. For calculation of the Tangential Sensitivity we will therefore assume that the peak value is $\sqrt{2}$ times the rms value.

The noise in the diode circuit is dependent on the diode construction and biasing techniques and will contain both thermal and shot noise components. A fairly accurate answer can be obtained by using a low noise amplifier at the output of the detector with an equivalent series noise voltage of, say, $3\text{nV}/\sqrt{\text{Hz}}$ and assuming that the diode contributes, say, a further $6\text{nV}/\sqrt{\text{Hz}}$, as described by Combes, Graffeuil and Sauntereau [1], producing a sum of the squares of around $7\text{nV}/\sqrt{\text{Hz}}$. The peak value would then be about $10\text{nV}/\sqrt{\text{Hz}}$.

Taking an ideal sensitivity for the detector of 1000mV/mW and a bandwidth of say 1 MHz, the tangential sensitivity is therefore around –50dBm.

1.12   Varactor Diodes

Varactor diodes have a voltage variable capacitance and are used to provide electronic tuning of oscillators and filters. For uniformly doped abrupt junction diodes, the capacitance–voltage (*C–V*) curve provides an inverse square root variation of capacitance with reverse voltage, V_R:

$$C(V) = \frac{C_0}{\sqrt{\left(1 + \dfrac{V_R}{\varphi}\right)}} \tag{1.175}$$

where C_0 is the zero bias capacitance, and φ is the built-in voltage, around 0.7V for silicon p–n junctions and GaAs Schottky barriers. Note that this equation applies for $V_R > 0$. As the resonant frequency of a tuned circuit is:

$$f_{res} = \frac{1}{2\pi\sqrt{LC}} \qquad (1.176)$$

This results in a tuning law which has a fourth root frequency dependence:

$$f_{res} = \frac{1}{2\pi\sqrt{LC_0}}\left(1 + \frac{V_R}{\varphi}\right)^{\frac{1}{4}} \qquad (1.177)$$

To obtain a linear variation of frequency with applied voltage an inverse square law is required for C - V as shown in equation (1.78).

$$C(V) = \frac{C_0}{\left(1 + \dfrac{V_R}{\varphi}\right)^2} \qquad (1.178)$$

thereby producing:

$$f_{res} = \frac{1}{2\pi\sqrt{LC_0}}\left(1 + \frac{V_R}{\varphi}\right) \qquad (1.179)$$

This can usually be achieved over a limited frequency range using hyperabrupt junction varactors.

As varactors operate in reverse bias, care must be taken to avoid breakdown. The reverse breakdown voltage is typically around 20V for silicon varactors.

The effect of the Q and series loss resistance of varactor diodes on the phase noise in oscillators is described in Chapter 4 on low noise oscillators.

1.13   Passive Components

In circuit design it is important to ensure that the value of components is fairly constant as the frequency is varied or that the variation is well understood. In resistors and capacitors there is inherent inductance due to the leads, size and shape (where typically lead inductance is around 1nH/mm) due to the energy stored in the magnetic field. Similarly there is parasitic capacitance both internally and also to the surrounding structure such as the printed circuit board.

1.13.1   Resistors

The equivalent circuit for a resistor is shown in Figure 1.28. This consists of a parasitic series inductance up to around 5 to 10nH and parallel capacitance up to a few picofarads. This is of course very dependent on the shape and size of the structure as well as the mounting method to the printed circuit board. Here it is worth highlighting chip resistors using information from data sheets.

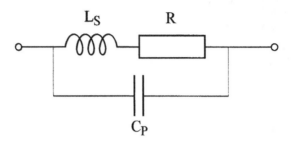

Figure 1.28 Equivalent circuit model for a resistor

Low power chip resistors are currently available in four sizes, 1206, 0805, 0603, 0402, where the first two digits are the length in 10s of thou and the last two digits are the width in 10s of thou. Therefore a 1206 resistor is 120 thou long and 60 thou wide. The values in mm are shown in Table 1,2 as well as typical values for parisitic capacitance and inductance collated from data sheets. Tolerances are not included. Note that these values are dependent on the track widths and angle of attachment. For example the inductance is likely to increase if the direction of current flow is changed suddenly. A few examples of the change in impedance with frequency for differing values of parasitics are included on the following pages.

Table 1.2 Chip resistors vs size and typical parasitic C and L

Resistor size	Length	Width	Capacitance (typ)	Inductance (typ)
1206	3.2mm	1.6mm	0.05pF	2nH
0805	2mm	1.25mm	0.09pF	1nH
0603	1.6mm	0.8mm	0.05pF	0.4nH
0402	1mm	0.5mm		

The variation of impedance with frequency is very important and typical variations of $|Z|/R$ are shown in Figure 1.29 assuming $C = 0.09$pF and $L = 1$nH. The phase shift is shown in Figure 1.30.

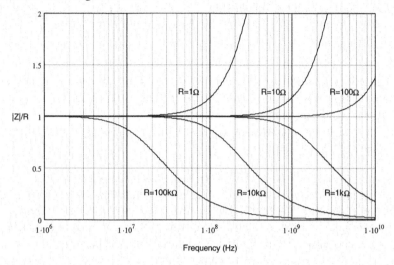

Figure 1.29 Variation of $|Z|/R$ with frequency

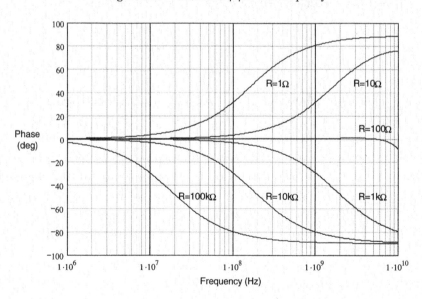

Figure 1.30 Typical variation of phase shift with frequency

It is interesting to note that the author has built some low cost 50Ω stripline circuits which required a reasonable 50Ω termination at 10.5GHz. The return loss using a 50Ω 1206 chip resistor was worse than 5dB. A return loss of better than 12dB was obtained with two ,0402, 100Ω chip resistors in parallel.

1.13.2 Capacitors

The equivalent circuit for capacitors is shown in Figure 1.31. Here there are series and parallel loss resistors and a series inductance. The series inductance will cause a series resonance at higher frequencies producing an effective decrease in impedance and hence an **increase** in effective capacitance as this frequency is approached from lower frequencies. The effect of series resonance is illustrated in Figure 1.32 and Figure 1.33 where the series inductance L_s is assumed to be 1nH. The series resistance for the 1000pF, 100pF and 10pF capacitors is assumed to be 0.08Ω, 0.2Ω and 0.5Ω respectively. The series resonant frequency vs capacitance is shown in figure 1.33 for three differing values of series inductance 0.6nH, 1nH, 1.5nH. R_p is ignored in these simulations. Note that if a large value capacitor is chosen for decoupling then its impedance may be higher than realised. For example, 1000pf chip capacitors resonate at around 200MHz so that above this frequency it is worth using a smaller C. It is useful to know that an empirical rule of thumb for lead inductance on conventional components is around 1nH/mm.

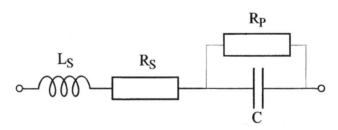

Figure 1.31 Equivalent circuit model for a capacitor

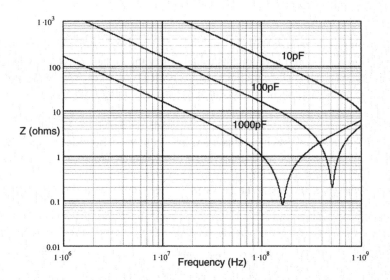

Figure 1.32 Effect of series resonance on Z

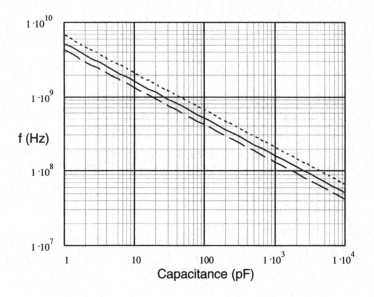

Figure 1.33 Series resonant frequency

1.13.3   Inductors

Inductors are often required in RF circuits for filtering, biasing and impedance transformation. Air cored components usually offer the highest Q typically ranging from 100 to 300, and surface mount versions offer Qs of 50 to 100. A non contact method for measuring Q is shown in Section 4.13 at the end of Chapter 4 on oscillators.

The equivalent circuit model for an inductor is shown in Figure 1.34 and consists of series and parallel loss resistors and, very importantly, a parallel capacitor which models the capacitance between windings. This therefore increases the impedance close to resonance giving a higher apparent inductance as resonance is approached (from the lower frequency side).

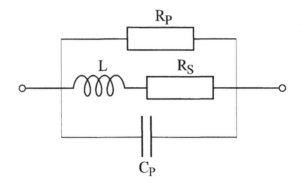

Figure 1.34 Equivalent circuit of an inductance of value L

An approximate formula for inductance is given by Vizmuller [9]. This formula neglects the wire diameter and is valid for coil length > 0.4 of the coil diameter.

$$L \approx \frac{(\text{coil diameter})^2 . N^2}{(0.45 \text{ coil diameter}) + (\text{coil length})} \qquad (1.180)$$

where L is the inductance in nH, and the lengths are in mm. It is interesting to perform a simple electromagnetic calculation for inductance based on Ampère's Law. Calculate the inductance of a long, thin, toroidal solenoid as shown in Figure 1.35. This structure is chosen as the symmetrical approximation for Ampère's Law is easily applied. The answer for a straight coil can then be obtained simply by breaking the toroid and assuming that the length is significantly larger than the width.

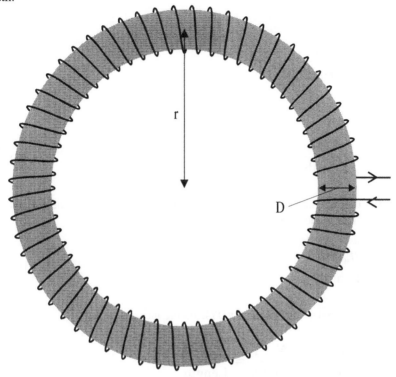

Figure 1.35 Toroidal Inductor

The solenoid has a number of turns, N, a radius, r, and coil diameter D. Inductance is defined as:

$$L = \frac{\Lambda}{I} = \left(\frac{\text{Flux Linkage}}{\text{Current}} \right) \qquad (1.181)$$

Taking Ampère's Law:

$$\oint H.dl = NI \qquad (1.182)$$

Apply Ampère's Law around the loop at radius r. As symmetry exists:

$$H.2\pi r = NI \tag{1.183}$$

Therefore:

$$H = \frac{NI}{2\pi r} = \frac{NI}{\text{Length}} \tag{1.184}$$

As:

$$B = \mu_0 H = \frac{\mu_0 NI}{\text{Length}} \tag{1.185}$$

Assuming that the flux density, B, does not vary over the radius inside the coil (as $r \ll D$), the flux is therefore:

$$\phi = BA = \frac{\mu_0 NIA}{\text{Length}} \tag{1.186}$$

where A is the area of the solenoid $\pi D^2/4$. As the inductance is:

$$L = \frac{\Lambda}{I} = \frac{N\phi}{I} = \frac{\mu_0 N^2 IA}{I.\text{Length}} = \frac{\mu_0 N^2 A}{\text{Length}} \tag{1.187}$$

and as:

$$\mu_0 = 4\pi \times 10^7 \text{ H/m} \tag{1.188}$$

The inductance in nH when the lengths are in mm is therefore:

$$L \approx \frac{N^2 (\text{coil diameter})^2}{(\text{coil length})} \tag{1.189}$$

which is very similar to the formula in equation (1.180) if the coil is assumed to be much longer than its diameter.

Ferrite cores with high permeability can be used to increase the inductance. If a closed core is used (which is often the case at lower frequencies) then great care should be taken not to saturate the core. The effect of saturation can be reduced by incorporating air gaps to reduce the flux density.

1.13.3.1 Spiral Inductor

Spiral inductors offer the potential of moderate Q with high reproducibility and low cost of manufacture. An approximate formula for the inductance of a spiral is given by Wheeler [10]:

$$L \approx \frac{0.394a^2 N^2}{(8a+11c)} \qquad (1.190)$$

where a is the mean radius = 0.5(outer radius + inner radius), N is the number of turns, and c is the radial depth of the winding (outer radius - inner radius).

A set of spiral inductors, made on 1/16th thickness FR4, both using the same spiral but with different scaling is shown in Figure 1.36. These were made by the author and consist of a set of small coils with three, four and five turns called 3S, 4S and 5S. A larger set are called 3L, 4L, 5L. All the coils had Qs greater than 100. Table 1.3 contains the measured inductance and Q for these coils at 120MHz and 130MHz using the non-contact Q measurement technique shown in section 4.13 in Chapter 4 on low noise oscillators. Also included in this data is an attempt to fit an $L = KN^2$ curve and it is shown that a reasonable fit can be obtained for each set of coils. Note that the turns are concentrated at maximum diameter to maintain high flux linkage (and hence inductance) with reduced loss.

Table 1.3 Inductance and Q of spiral inductors

Coils	Inductance, nH	Q at 120MHz	Q at 130 MHz	$L = K.N^2$
5S	350	126	125	14
4S	240	135	124	15
3S	135	106	104	15
5L	509	110	129	20
4L	302	133	122	18.8
3L	167	119	125	18.5

14 cm

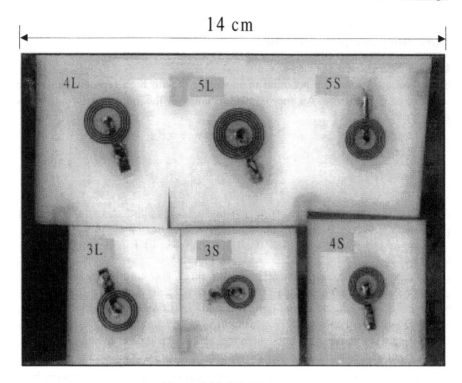

Figure 1.36 Spiral Inductors

1.14   References and Bibliography

1. Paul F. Combes, Jacques Graffeuil and J.-F. Sautereau, *Microwave Components, Devices and Active Circuit*, Wiley, 1987.
2. P.R. Gray and R.G. Meyer, *Analysis and Design of Analogue Integrated Circuits*, Wiley, 1993.
3. W.H. Haywood, "Introduction to Radio Frequency Design", Prentice Hall, 1982.
4. Hayward and DeMaw, *Solid State Design for the Radio Amateur*, US Amateur Radio Relay League.
5. Chris Bowick, *RF Circuit Design*, SAMS, Division of Macmillan, 1987.
6. H.L. Krauss, C.W. Bostian and F.H. Raab, *Solid State Radio Engineering*, Wiley, 1980.
7. Peter C.L. Yip, *High Frequency Circuit Design and Measurements*, Chapman and Hall, 1990.
8. *RF Wideband Transistors*, Product Selection 2000 Discrete Semiconductors CD (First Edition) Release 01-2000.
9. P. Vizmuller, *RF Design Guide, Systems, Circuits, and Equations*, Artech House 1995.
10. H.A. Wheeler, "Simple Inductance Formulas for Radio Coils", IRE Proceedings, 1928

2

Two Port Network Parameters

2.1 Introduction

This chapter will describe the important linear parameters which are currently used to characterise two port networks. These parameters enable manipulation and optimisation of RF circuits and lead to a number of figures of merit for devices and circuits. Commonly used figures of merit include h_{FE}, the short circuit low frequency current gain, f_T, the transition frequency at which the modulus of the short circuit current gain equals one, GUM (Maximum Unilateral Gain), the gain when the device is matched at the input and the output and the internal feedback has been assumed to be zero. All of these figures of merit give some information of device performance but the true worth of them can only be appreciated through an understanding of the boundary conditions defined by the parameter sets.

The most commonly used parameters are the z, y, h, $ABCD$ and S parameters. These parameters are used to describe linear networks fully and are interchangeable. Conversion between them is often used as an aid to circuit design when, for example, conversion enables easy deconvolution of certain parts of an equivalent circuit. This is because the terminating impedance's and driving sources vary. Further if components are added in parallel the admittance parameters can be directly added; similarly if they are added in series impedance parameters can be used. Matrix manipulation also enables easy conversion between, for example, common base, common emitter and common collector configurations.

For RF design the most commonly quoted parameters are the y, h and S parameters and within this book familiarity with all three parameters will be required for circuit design. For low frequency devices the h and y parameters are quoted. At higher frequencies the S parameters h_{FE} and f_T are usually quoted. It is often easier to obtain equivalent circuit information more directly from the h and y

parameters, however, the later part of the chapter will describe how S parameters can be deconvolved.

All these parameters are based on voltages, currents and travelling waves applied to a network. Each of them can be used to characterise linear networks fully and all show a generic form. This chapter will concentrate on two port networks though all the rules described can be extended to N port devices.

The z, y, h and $ABCD$ parameters cannot be accurately measured at higher frequencies because the required short and open circuit tests are difficult to achieve over a broad range of frequencies. The scattering (S) parameters are currently the easiest parameters to measure at frequencies above a few tens of MHz as they are measured with 50Ω or 75Ω network analysers. The network analyser is the basic measurement tool required for most RF and microwave circuit design and the modern instrument offers rapid measurement and high accuracy through a set of basic calibrations. The principle of operation will be described in the Chapter 3 on amplifier design (measurements section).

Note that all these parameters are linear parameters and are therefore regarded as being independent of signal power level. They can be used for large signal design over small perturbations but care must be taken. This will be illustrated in the Chapter 6 on power amplifier design.

A two port network is shown in Figure 2.1.

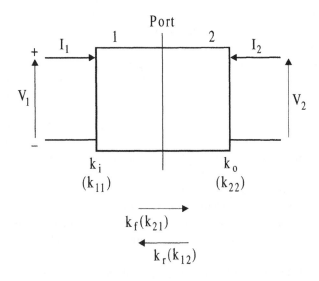

Figure.2.1 General representation of a two port network.

The first point to note is the direction of the currents. The direction of the current is into both ports of the networks. There is therefore symmetry about a central line. This is important as inversion of a symmetrical network must not change the answer. For a two port network there are four parameters which are measured:

k_{11} = the input (port 1) parameter

k_{22} = the output (port 2) parameter

k_{21} = the forward transfer function

k_{12} = the reverse transfer function

As mentioned earlier there is a generic form to all the parameters. This is most easily illustrated by taking the matrix form of the two port network and expressing it in terms of the dependent and independent variables.

Dependent   Parameters   Independent
variables   variables

$$\begin{pmatrix} \Phi d_1 \\ \Phi d_2 \end{pmatrix} = \begin{pmatrix} k_i & k_r \\ k_f & k_o \end{pmatrix} \begin{pmatrix} \Phi i_1 \\ \Phi i_2 \end{pmatrix} \tag{2.1}$$

In more normal notation:

$$\begin{pmatrix} \Phi d_1 \\ \Phi d_2 \end{pmatrix} = \begin{pmatrix} k_{11} & k_{12} \\ k_{21} & k_{22} \end{pmatrix} \begin{pmatrix} \Phi i_1 \\ \Phi i_2 \end{pmatrix} \tag{2.2}$$

Therefore:

$$\phi_{d1} = k_{11}\phi i_{i1} + k_{12}\phi_{i2} \tag{2.3}$$

$$\phi_{d2} = k_{21}\phi i_{i1} + k_{22}\phi_{i2} \tag{2.4}$$

One or other of the independent variables can be set to zero by placing a S/C on a port for the parameters using voltages as the independent variables, an O/C for the parameters using current as the independent variable and by placing Z_0 as a termination when dealing with travelling waves.

Therefore in summary:

CURRENTS SET TO ZERO BY TERMINATING IN AN O/C

VOLTAGES SET TO ZERO BY TERMINATING IN A S/C

REFLECTED WAVES SET TO ZERO BY TERMINATION IN Z_0

Now let us examine each of the parameters in turn.

2.2   Impedance Parameters

The current is the independent variable which is set to zero by using O/C terminations. These parameters are therefore called the O/C impedance parameters. These parameters are shown in the following equations:

$$\begin{pmatrix} V_1 \\ V_2 \end{pmatrix} = \begin{pmatrix} z_{11} & z_{12} \\ z_{21} & z_{22} \end{pmatrix} \begin{pmatrix} I_1 \\ I_2 \end{pmatrix}$$

(2.5)

$$V_1 = Z_{11}I_1 + Z_{12}I_2$$

(2.6)

$$V_2 = Z_{21}I_1 + Z_{22}I_2$$

(2.7)

$$z_{11} = \frac{V_1}{I_1}(I_2 = 0)$$

(2.8)

$$z_{12} = \frac{V_1}{I_2}(I_1 = 0)$$

(2.9)

$$z_{21} = \frac{V_2}{I_1}(I_2 = 0)$$

(2.10)

$$z_{22} = \frac{V_2}{I_2}(I_1 = 0)$$

(2.11)

z_{11} is the input impedance with the output port terminated in an O/C ($I_2 = 0$). This may be measured, for example, by placing a voltage V_1 across port 1 and measuring I_1.

Similarly z_{22} is the output impedance with the input terminals open circuited. z_{21} is the forward transfer impedance with the output terminal open circuited and z_{12} is the reverse transfer impedance with the input port terminated in an O/C.

Open circuits are not very easy to implement at higher frequencies owing to fringing capacitances and therefore these parameters were only ever measured at low frequencies. When measuring an active device a bias network was required. This should still present an O/C at the signal frequencies but of course should be a short circuit to the bias voltage. This would usually consist of a large inductor with a low series resistance.

A Thévenin equivalent circuit for the z parameters is shown in Figure 2.2. This is an abstract representation for a generic two port.

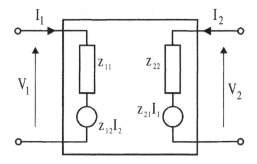

Figure 2.2 Thévenin equivalent circuit for z parameter model

The effect of a non-ideal O/C means that these parameters would produce most accurate results for measurements of fairly low impedances. Thus for example these parameters would be more accurate for the forward biased base emitter junction rather than the reverse biased collector base junction.

The open circuit parameters were used to some extent in the early days of transistor development at signal frequencies up to a few megahertz but with advances in technology they are now very rarely used in specification sheets. They are, however, useful for circuit manipulation and have a historical significance.

Now let us look at the S/C y parameters where the voltages are the independent variables. These are therefore called the S/C admittance parameters and describe the input, output, forward and reverse admittances with the opposite port terminated in a S/C. These parameters are regularly used to describe FETs and dual gate MOSFETs up to 1 GHz and we shall use them in the design of VHF amplifiers. To enable simultaneous measurement and biasing of a network at the measurement frequency large capacitances would be used to create the S/C. Therefore for accurate measurement the effect of an imperfect S/C means that these parameters are most accurate for higher impedance networks. At a single frequency a transmission line stub could be used but this would need to be retuned for every different measurement frequency.

2.3   Admittance Parameters

V_1 and V_2 are the independent variables. These are therefore often called S/C y parameters. They are often useful for measuring higher impedance circuits, i.e. they are good for reverse biased collector base junctions, but less good for forward biased base emitter junctions. For active circuits a capacitor should be used as the load.

The y parameter matrix for a two port is therefore:

$$\begin{pmatrix} I_1 \\ I_2 \end{pmatrix} = \begin{pmatrix} y_{11} & y_{12} \\ y_{21} & y_{22} \end{pmatrix} \begin{pmatrix} V_1 \\ V_2 \end{pmatrix}$$

(2.12)

$$I_1 = y_{11}V_1 + y_{12}V_2$$

(2.13)

$$I_2 = y_{21}V_1 + y_{22}V_2$$

(2.14)

The input admittance with the output S/C is:

$$y_{11} = \frac{I_1}{V_1} \quad (V_2 = 0)$$

(2.15)

The output admittance with the input S/C is:

$$y_{22} = \frac{I_2}{V_2} \quad (V_1 = 0)$$

(2.16)

The forward transfer admittance with the output S/C is:

$$y_{21} = \frac{I_2}{V_1} \quad (V_2 = 0) \tag{2.17}$$

The reverse transfer admittance with the input S/C is:

$$y_{12} = \frac{I_1}{V_2} \quad (V_1 = 0) \tag{2.18}$$

A Norton equivalent circuit model for the y parameters is shown in Figure 2.3.

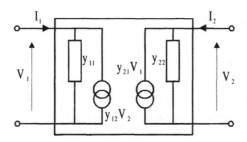

Figure 2.3 Norton equivalent circuit for y parameter model

It is often useful to develop accurate, large signal models for the active device when designing power amplifiers. An example of a use that the author has made of y parameters is shown here. It was necessary to develop a non-linear model for a 15 watt power MOSFET to aid the design of a power amplifier. This was achieved by parameter conversion to deduce individual component values within the model. If we assume that the simple low to medium frequency model for a power FET can be represented as the equivalent circuit shown in Figure 2.4,

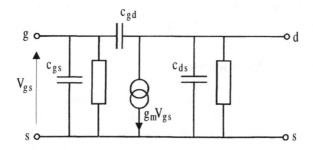

Figure 2.4 Simple model for Power FET

then to obtain the π capacitor network the S parameters were measured at different bias voltages. These were then converted to y parameters enabling the three capacitors to be deduced. The non-linear variation of these components with bias could then be derived and modelled. The measurements were taken at low frequencies (50 to 100MHz) to ensure that the effect of the parasitic package inductances could be ignored. The equations showing the relationships between the y parameters and the capacitor values are shown below. This technique is described in greater detail in Chapter 6 on power amplifier design.

$$\mathrm{Im}\, y_{11} = \left(C_{gs} + C_{gd} \right) \omega \qquad\qquad (2.19)$$

$$\mathrm{Im}\, y_{22} = \left(C_{ds} + C_{gd} \right) \omega \qquad\qquad (2.20)$$

$$\mathrm{Im}\, y_{12} = - \left(C_{gd} \right) \omega \qquad\qquad (2.21)$$

where Im refers to the imaginary part.

Note that this form of parameter conversion is often useful in deducing individual parts of a model where an O/C or S/C termination enables different parts of the model to be deduced more easily.

It has been shown that an O/C can be most accurately measured when terminated in a low impedance and that low impedances can be most accurately measured in a high impedance load.

If the device to be measured has a low input impedance and high output impedance then a low output impedance termination and a high input termination are required. To obtain these the Hybrid parameters were developed. In these parameters V_2 and I_1 are the independent variables. These parameters are used to describe the Hybrid π model for the Bipolar Transistor. Using these parameters two figures of merit, very useful for LF, RF and Microwave transistors have been developed. These are h_{fe} which is the Low frequency short circuit current gain and f_T which is called the transition Frequency and occurs when the Modulus of the Short circuit current gain is equal to one.

2.4   Hybrid Parameters

If the circuit to be measured has a fairly low input impedance and a fairly high output impedance as in the case of common emitter or common base configurations, we require the following for greatest accuracy of measurement: A S/C at the output so V_2 is the independent variable and an open circuit on the input so I_1 is the independent variable. Therefore:

$$
\begin{pmatrix} V_1 \\ I_2 \end{pmatrix} = \begin{pmatrix} h_{11} & h_{12} \\ h_{21} & h_{22} \end{pmatrix} \begin{pmatrix} I_1 \\ V_2 \end{pmatrix}
\tag{2.22}
$$

$$
V_1 = h_{11}I_1 + h_{12}V_2
\tag{2.23}
$$

$$
I_2 = h_{21}I_1 + h_{22}V_2
\tag{2.24}
$$

$$
h_{11} = \frac{V_1}{I_1} \quad (V_2 = 0, \text{S/C})
\tag{2.25}
$$

$$
h_{22} = \frac{I_2}{V_2} \quad (I_1 = 0, \text{O/C})
\tag{2.26}
$$

$$
h_{21} = \frac{I_2}{I_1} \quad (V_2 = 0, \text{S/C})
\tag{2.27}
$$

$$
h_{12} = \frac{V_1}{V_2}(I_1 = 0, \text{O/C})
\tag{2.28}
$$

Therefore h_{11} is the input impedance with the output short circuited. h_{22} is the output admittance with the input open circuited. h_{21} is the S/C current gain (output = S/C) and h_{12} is the reverse voltage transfer characteristic with the input open circuited.

Note that these parameters have different dimensions hence the title 'Hybrid Parameters'. Two often quoted and useful figures of merit are:

h_{fe} is the LF S/C current gain: h_{21} as $\omega \rightarrow 0$

f_T is the frequency at which $|h_{21}| = 1$. This is calculated from measurements made at a much lower frequency and then extrapolated along a $1/f$ curve.

2.5   Parameter Conversions

For circuit manipulation it is often convenient to convert between parameters to enable direct addition. For example, if you wish to add components in series, the parameter set can be converted to z parameters, and then added (Figure 2.5). Similarly if components are added in parallel then the y parameters could be used (Figure 2.6).

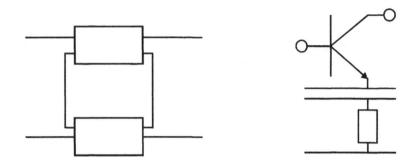

Figure 2.5 Illustration of components added in series

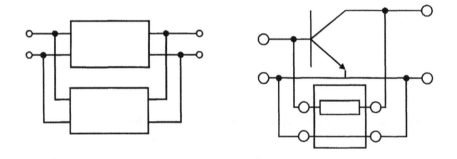

Figure 2.6 Illustration of components added in parallel

The *ABCD* parameters can be used for cascade connections. Note that they relate the input voltage to the output voltage and the input current to the negative of the output current. This means that they are just multiplied for cascade connections as the output parameters become the input parameters for the next stage.

$$\begin{pmatrix} V_1 \\ I_1 \end{pmatrix} = \begin{pmatrix} A & B \\ C & D \end{pmatrix} \begin{pmatrix} V_2 \\ -I_2 \end{pmatrix}$$

(2.29)

$$V_1 = AV_2 - BI_2 \qquad\qquad\qquad (2.30)$$

$$I_1 = CV_2 - DI_2 \qquad\qquad\qquad (2.31)$$

2.6   Travelling Wave and Scattering Parameters

Accurate open and short circuits are very difficult to produce over broad and high frequency ranges owing to parasitic effects. Devices are also often unstable when loaded with an O/C or S/C and the biasing requirements also add problems when O/C and S/C loads are used. The effect of the interconnecting leads between the test equipment and the device under test (DUT) also becomes critical as the frequencies are increased.

For this reason the scattering parameters (S parameters) were developed and these are based on voltage travelling waves normalised to an impedance such that when squared they become a power. They relate the forward and reverse travelling waves incident on a network. Before the S parameters are considered, the propagation of waves in transmission lines will be reviewed and the concept of reflection coefficient for a one port network will be discussed.

2.6.1   Revision of Transmission Lines

The notation that is used is quite important here and we shall use the symbol V^+ to represent the forward wave and V^- as the reverse wave such that these waves are described by the following equations:

$$V^+ = \mathrm{Re}\{A\exp[j(\omega t - \beta z)]\} \quad \text{FORWARD WAVE} \qquad (2.32)$$

$$V^- = \mathrm{Re}\{A\exp[j(\omega t + \beta z)]\} \quad \text{REVERSE WAVE} \qquad (2.33)$$

The forward wave V^+ is the real part of the exponential which is a sinusoidal travelling wave. These waves show a linear phase variation of similar form in both time and space. Hence the phase changes with time owing to the frequency ωt and with space due to the propagation coefficient βz. Note that by convention the forward wave is $-\beta z$ whereas the reverse wave is $+\beta z$.

It is important to know the voltage and the current at any point along a transmission line. Here the voltage at a point is just the sum of the voltages of the forward and reverse waves:

$$V(x,t) = V^+ + V^-$$
(2,34)

Similarly the currents are also summed; however, note that we define the direction of the current as being in the forward direction. The sum of the currents is therefore the subtraction of the magnitude of the forward and reverse currents:

$$I(x,t) = I^+ + I^- = \frac{V^+}{Z_o} - \frac{V^-}{Z_o}$$
(2.35)

From these definitions we can now derive expressions for the reflection coefficient of a load Z_L at the end of a transmission line of characteristic impedance Z_0 in terms of Z_L and Z_o. This is illustrated in Figure 2.7 where the reflection coefficient, ρ, is the ratio of the reverse wave to the forward wave:

$$\rho = \frac{V^-}{V^+}$$
(2.36)

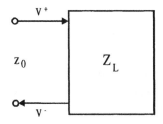

Figure 2.7 Reflection coefficient of a load Z_L

Remember that:

$$V_{in} = V^+ + V^-$$
(2.37)

$$I_{in} = I^+ + I^- = \frac{V^+}{Z_o} - \frac{V^-}{Z_o}$$
(2.38)

The impedance is therefore the ratio of the total voltage to the total current at Z_L:

$$\frac{V_T}{I_T} = \frac{V^+ + V^-}{\dfrac{V^+}{Z_o} - \dfrac{V^-}{Z_o}} = Z_L \qquad (2.39)$$

If we divide by V^+ to normalise the equation to the incident wave then:

$$\frac{1 + \dfrac{V^-}{V^+}}{\dfrac{1}{Z_o} - \dfrac{V^-}{V^+ Z_o}} = Z_L \qquad (2.40)$$

as

$$\frac{V^-}{V^+} = \rho \qquad (2.36)$$

then:

$$1 + \rho = \frac{Z_L}{Z_o}\left[1 - \rho\right] \qquad (2.41)$$

$$\rho\left[1 + \frac{Z_L}{Z_o}\right] = \frac{Z_L}{Z_o} - 1 \qquad (2.42)$$

$$\rho = \frac{Z_L - Z_o}{Z_L + Z_o} \qquad (2.43)$$

Note also that:

$$Z_L = Z_0 \frac{1 + \rho}{1 - \rho} \qquad (2.44)$$

If $Z_L = Z_0$, $\rho = 0$ as there is no reflected wave, In other words, all the power is absorbed in the load. If $Z_L = O/C$, $\rho = 1$ and if $Z_L = 0$, $\rho = -1$ (i.e. $V^- = -V^+$).

The voltage and current wave equations along a transmission line will be determined to enable the calculation of the characteristic impedance of a transmission line and to calculate the variation of impedance along a line when the line is terminated in an arbitrary impedance.

2.6.2   Transmission Lines (Circuit Approach)

Transmission lines are fully distributed circuits with important parameters such as inductance per unit length, capacitance per unit length, velocity and characteristic impedance. At RF frequencies the effect of higher order transverse and longitudinal modes can be ignored for cables where the diameter is less than $\lambda/10$ and therefore such cables can be modelled as cascaded sections of short elements of inductance and capacitance. The variations of voltage and current along the line obey the standard circuit equations for voltages and currents in inductors and capacitors of:

$$V = L\, dI/dt \tag{2.45}$$

and:

$$V = \frac{1}{C}\int I\, dt \tag{2.46}$$

When dealing with transmission lines they are expressed as partial derivatives because both the voltages and currents vary in both time and space (equations 2.47 and 2.48). Models for a transmission line are shown in Figure 2.8.

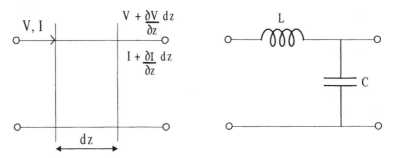

Figure 2.8 Model for a transmission line

$$\frac{\partial v}{\partial z} dz = -Ldz \frac{\partial I}{\partial t}$$ (2.47)

$$\frac{\partial v}{\partial z} = -L \frac{\partial I}{\partial t}$$ (2.48)

$$\frac{\partial I}{\partial z} dz = -Cdz \frac{\partial v}{\partial t}$$ (2.49)

$$\frac{\partial I}{\partial z} = -C \frac{\partial v}{\partial t}$$ (2.50)

Differentiating (2.48) with respect to t gives:

$$\frac{\partial^2 v}{\partial z \partial t} = -L \frac{\partial^2 I}{\delta t^2}$$ (2.51)

Differentiating (2.50) with respect to z gives:

$$\frac{\partial^2 I}{\partial z^2} = -C \frac{\partial^2 v}{\partial t \partial z}$$ (2.52)

As the order of differentiation is unimportant, substitute (2.51) in (2.52) then:

$$\frac{\partial^2 I}{\partial z^2} = LC \frac{\partial^2 I}{\partial t^2}$$ (2.53)

Similarly:

$$\frac{\partial^2 v}{\partial z^2} = -L \frac{\partial^2 I}{\partial t \partial z}$$ (2.54)

$$\frac{\partial^2 I}{\partial z \partial t} = -C \frac{\partial^2 v}{\partial t^2}$$ (2.55)

$$\frac{\partial^2 v}{\partial z^2} = LC\frac{\partial^2 v}{\partial t^2}$$

(2.56)

The solutions to these equations are wave equations of standard form where the velocity, υ, is:

$$\upsilon = \frac{1}{\sqrt{LC}}$$

(2.57)

General solutions are in the form of a forward and reverse wave:

Forward wave   Reverse wave

$$V = F_1\left(t - \frac{z}{\upsilon}\right) + F_2\left(t + \frac{z}{\upsilon}\right)$$

(2.58)

The usual solution is sinusoidal in form:

$$V = V_f\, e^{j(\omega t - \beta z)} + V_r\, e^{j(\omega t + \beta z)}$$

(2.59)

where β is the propagation coefficient:

$$\beta = \frac{2\pi}{\lambda} = \frac{2\pi f}{\upsilon} = \omega\sqrt{LC}$$

(2.60)

To calculate the current, take equation (2.48)

$$\frac{\partial v}{\partial z} = -L\frac{\partial I}{\partial t}$$

(2.48)

Substitute (2.59) in (2.48)

$$-L\frac{\partial I}{\partial t} = \frac{\partial v}{\partial z} = -\beta V_f\, e^{j(\omega t - \beta z)} + \beta V_r\, e^{j(\omega t + \beta z)}$$

(2.61)

and integrate with respect to t:

$$I = \frac{\beta}{L}\left[\frac{1}{\omega}V_f\, e^{j(\omega t - \beta z)} - \frac{1}{\omega}V_r\, e^{j(\omega t + \beta z)}\right] \qquad (2.62)$$

as:

$$\beta = \frac{2\pi f}{\upsilon} \qquad (2.63)$$

then:

$$I = \frac{1}{L\upsilon}\left[V_f\, e^{j(\omega t + \beta z)} - V_r\, e^{j(\omega t + \beta z)}\right] \qquad (2.64)$$

2.6.3   Characteristic Impedance

This is an important parameter for transmission lines and is the impedance that would be seen if a voltage was applied to an infinite length of lossless line. Further when this line is terminated in this impedance the impedance is independent of length as no reverse wave is produced ($\rho = 0$). The characteristic impedance is therefore:

$$\frac{V^+}{I^+}\quad \text{or}\quad \frac{V^-}{I^-} = Z_o$$

as before V^+ denotes the forward wave and V^- denotes the reverse wave. Remember that:

$$V^+ = V_f\, e^{j(\omega t - \beta z)} \qquad (2.65)$$

$$I^+ = I = \frac{\beta}{L}\left[\frac{1}{\omega}V_f\, e^{j(\omega t - \beta z)}\right] = \frac{1}{L\upsilon}\left[V_f\, e^{f(\omega t - \beta z)}\right] \qquad (2.66)$$

Therefore the characteristic impedance of the line is:

$$\frac{V^+}{I^+} = L\upsilon = Z_0 \qquad (2.67)$$

and as:

$$v = \frac{1}{\sqrt{LC}} \qquad\qquad (2.68)$$

$$Z_o = \sqrt{\frac{L}{C}} \qquad\qquad (2.69)$$

Note that the impedance along a line varies if there is both a forward and reverse wave due to the phase variation between the forward and reverse waves.

$$Z = \frac{V_T}{I_T} = \frac{V^+ + V^-}{I^+ - I^-} \qquad\qquad (2.70)$$

2.6.4   Impedance Along a Line Not Terminated in Z_0

When there is both a forward and reverse wave within a transmission line at the same time the input impedance varies along the line as $Z_{in} = V_T/I_T$. Consider the impedance along a transmission line of impedance Z_0 of length L terminated in an arbitrary impedance Z_L (Figure 2.9).

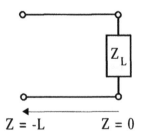

Figure 2.9 Impedance variation along a transmission line not terminated in Z_0

At $Z = 0$, let $V^+ = V_f e^{j\omega t}$ and $V^- = \rho V^+$ Then Z_{in} at the input of the line at the point where $Z = -L$ is:

$$Z_{in} = \frac{V}{I} = Z_o \left[\frac{V^+ e^{j\beta L} + V^- e^{-j\beta L}}{V^+ e^{j\beta L} - V^- e^{-j\beta L}} \right] \qquad (2.71)$$

$$Z_{in} = Z_o \frac{e^{j\beta L} + \rho e^{-j\beta L}}{e^{j\beta L} - \rho e^{-j\beta L}} \qquad (2.72)$$

As:

$$\rho = \frac{Z_L - Z_0}{Z_L + Z_0} \qquad (2.73)$$

$$Z_{in} = Z_o \frac{e^{j\beta L} + \rho e^{-j\beta L}}{e^{j\beta L} - \rho e^{-j\beta L}} \qquad (2.74)$$

As:

$$\cos\theta = \frac{e^{j\theta} + e^{-j\theta}}{2} \qquad (2.75)$$

and:

$$\sin\theta = \frac{e^{j\theta} - e^{-j\theta}}{2j} \qquad (2.76)$$

$$Z_{in} = Z_o \left[\frac{Z_L \cos\beta L + jZ_o \sin\beta L}{Z_o \cos\beta L + jZ_L \sin\beta L} \right] \qquad (2.77)$$

$$Z_{in} = Z_o \left[\frac{Z_L + jZ_o \tan\beta L}{Z_o + jZ_L \tan\beta L} \right] \qquad (2.78)$$

2.6.5   Non Ideal Lines

In this case:

$$Z_o = \sqrt{\frac{R + j\omega L}{G + j\omega C}}$$ (2.79)

and:

$$Z_{in} = Z_o \left[\frac{Z_L \cosh \gamma L + Z_o \sinh \gamma L}{Z_o \cosh \gamma L + Z_L \sinh \gamma L} \right]$$ (2.80)

where the propagation coefficient is:

$$\gamma = \alpha + j\beta = \sqrt{(R + j\omega L)(G + j\omega C)}$$ (2.81)

2.6.6   Standing Wave Ratio (SWR)

This is the ratio of the maximum AC voltage to the minimum AC voltage on the line and is often quoted at the same time as the return loss of the load.

For a line terminated in Z_0 the *SWR* is one. For an O/C or S/C the SWR is infinite as V_{min} is zero.

$$SWR = \frac{V_{max}}{V_{min}} = \frac{|V^+| + |V^-|}{|V^+| - |V^-|} = \frac{1 + |\rho|}{1 - |\rho|}$$ (2.82)

The impedance along a transmission line often used to be obtained by measuring both the magnitude and phase of V_{total} using a probe inserted into a slotted transmission line. The probe usually consisted of a diode detector operating in the square law region as described in Chapter 1 on device models.

2.7   Scattering Parameters

As mentioned earlier, accurate open and short circuits are very difficult to produce over broad and high frequency ranges owing to parasitic effects. Devices are also often unstable when loaded with an O/C or S/C and the biasing requirements also add problems when O/C and S/C loads are used. The use of 50Ω or 75Ω impedances solves most of these problems. The effect of the interconnecting leads between the test equipment and the device under test (DUT) also becomes critical as the frequencies are increased.

For this reason the scattering parameters (S parameters) were developed. These are based on voltage travelling waves normalised to an impedance such that when squared they become a power. They relate the forward and reverse travelling waves incident on a network

Further by using coaxial cables of the same impedance as the network analyser (typically 50 or 75Ω), the effects of the interconnecting leads can easily be included and for high accuracy measurements, error correction can be used. This will be described within the measurements section in Chapter 3 on amplifier design.

When testing port 1, an incident travelling wave is applied to port 1 and the output is terminated in Z_0. This means that there is no reflected wave from Z_{02} re-incident on port 2.

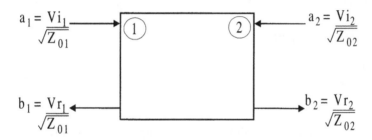

Figure 2.10 Two port model for S parameters

The input and reflected waves can be thought of as voltage travelling waves incident and reflected from a port normalised to the port impedance Z_{0n} such that when squared the wave is equal to the power travelling along the line. This is illustrated in Figure 2.10. These waves are defined in terms of:

a_n: incident waves on port n and

b_n: reflected waves from port n.

For a two port network the forward and reverse waves are therefore defined as:

Incident wave on port 1:

$$a_1 = \frac{Vi_1}{\sqrt{Z_{01}}}$$
(2.83)

Reflected wave from port 1:

$$b_1 = \frac{Vr_1}{\sqrt{Z_{01}}}$$
(2.84)

Incident wave on port 2:

$$a_2 = \frac{Vi_2}{\sqrt{Z_{02}}}$$
(2.85)

Reflected wave from port 2:

$$b_2 = \frac{Vr_2}{\sqrt{Z_{02}}}$$
(2.86)

The independent variables in this case are the input travelling waves on the port not being tested. These are made zero by terminating the ports with the characteristic impedance defined for the measurements called Z_{0n}. For example S_{11} is the ratio of the reflected power to the incident power at port 1 when there is no power incident on port 2 because port 2 has been terminated in Z_{02}. Z_0 is typically 50 or 75Ω. The S parameters can therefore be expressed in matrix form as:

Dependent   Independent
variables   variables

$$\begin{pmatrix} b_1 \\ b_2 \end{pmatrix} = \begin{pmatrix} S_{11} & S_{12} \\ S_{21} & S_{22} \end{pmatrix} \begin{pmatrix} a_1 \\ a_2 \end{pmatrix}$$
(2.87)

when expanded:

$$b_1 = S_{11}a_1 + S_{12}a_2$$
(2.88)

$$b_2 = S_{21}a_1 + S_{22}a_2 \qquad\qquad (2.89)$$

The input reflection coefficient with the output terminated in Z_{02} is therefore:

$$S_{11} = \frac{b_1}{a_1}\bigg|a_2 = 0 \qquad\qquad (2.90)$$

The forward transmission coefficient with the output terminated in Z_{02} is therefore:

$$S_{21} = \frac{b_2}{a_1}\bigg|a_2 = 0 \qquad\qquad (2.91)$$

The output reflection coefficient with the input terminated in Z_{01} is therefore:

$$S_{22} = \frac{b_2}{a_2}\bigg|a_1 = 0 \qquad\qquad (2.92)$$

The reverse transmission coefficient with the input terminated in Z_{01} is therefore:

$$S_{12} = \frac{b_1}{a_2}\bigg|a_1 = 0 \qquad\qquad (2.93)$$

To obtain circuit information from the S parameters it is necessary to calculate the S parameters in terms of V_{out}/V_{in}.

$$S_{21} = \frac{b_2}{a_1} = \frac{Vr_2}{\sqrt{Z_{02}}} \times \frac{\sqrt{Z_{01}}}{Vi_1} \qquad\qquad (2.94)$$

Note that:

$$\frac{V_{in}}{\sqrt{Z_{01}}} = a_1 + b_1 \qquad\qquad (2.95)$$

and that:

$$\frac{V_{out}}{\sqrt{Z_{02}}} = b_2 \tag{2.96}$$

Therefore:

$$\frac{\sqrt{Z_{01}}}{\sqrt{Z_{02}}} \times \frac{V_{out}}{V_{in}} = \left(\frac{b_2}{a_1 + b_1} \right) \tag{2.97}$$

$$\frac{\sqrt{Z_{01}}}{\sqrt{Z_{02}}} \times (a_1 + b_1) \times \frac{V_{out}}{V_{in}} = (b_2) \tag{2.98}$$

Dividing throughout by a_1:

$$\frac{\sqrt{Z_{01}}}{\sqrt{Z_{02}}} \times \left(\frac{a_1 + b_1}{a_1} \right) \times \frac{V_{out}}{V_{in}} = \left(\frac{b_2}{a_1} \right) = S_{21} \tag{2.99}$$

$$S_{21} = \frac{\sqrt{Z_{01}}}{\sqrt{Z_{02}}} \times \frac{V_{out}}{V_{in}} (1 + S_{11}) \tag{2.100}$$

Note that in most instances:

$$\frac{\sqrt{Z_{01}}}{\sqrt{Z_{02}}} = 1 \tag{2.101}$$

as the source and load impedances are the same, therefore:

$$S_{21} = \frac{V_{out}}{V_{in}} (1 + S_{11}) \tag{2.102}$$

If S_{11} was zero (in other words, if $Z_{in} = Z_{01}$) then $S_{21} = V_{out}/V_{in}$. S_{12} would be calculated by turning the network around, terminating it in Z_0 and calculating S_{21} as before.

2.7.1   Example Calculation Using S Parameters

This will now be illustrated by calculating the S parameters for a resistive network. It will be shown after this calculation that there is a simplification that can be made for calculating the forward and reverse parameters.

We calculate the S parameters for the resistive network consisting of Z_1 and Z_2 as shown in Figure 2.11, where the characteristic impedance for both ports is Z_0.

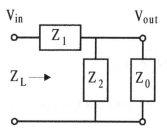

Figure 2.11 Network for example calculation of S paramters

To calculate S_{11} (ρ) the input impedance Z_L is required with the network terminated in Z_0.

$$Z_L = Z_2 // Z_o + Z_1 = \frac{Z_2 Z_o + Z_1 Z_2 + Z_1 Z_o}{Z_2 + Z_o} \tag{2.103}$$

$$S_{11} = \frac{Z_L - Z_o}{Z_L + Z_o} = \frac{Z_1 Z_2 + Z_1 Z_o - Z_o^{\,2}}{Z_1 Z_2 + Z_1 Z_o + 2 Z_2 Z_o + Z_o^{\,2}} \tag{2.104}$$

To calculate S_{21} it is now necessary to calculate V_{out}/V_{in}:

$$\frac{V_{out}}{V_{in}} = \frac{Z_2 // Z_o}{Z_2 // Z_o + Z_1} = \frac{Z_2 Z_o}{Z_2 Z_o + Z_2 Z_1 + Z_1 Z_o} \tag{2.105}$$

as:

$$S_{21} = \frac{V_{out}}{V_{in}} \left(1 + S_{11}\right) \tag{2.102}$$

$$S_{21} = \frac{Z_2 Z_o}{Z_1 Z_2 + Z_1 Z_o + Z_2 Z_o} \left[1 + \frac{Z_1 Z_2 + Z_1 Z_0 - Z_0^2}{Z_1 Z_2 + Z_1 Z_0 + 2Z_2 + Z_0^2} \right] \qquad (2.106)$$

$$S_{21} = \frac{Z_2 Z_o \left(2Z_1 Z_2 + 2Z_{11} Z_o + 2Z_2 Z_o \right)}{\left(Z_1 Z_2 + Z_1 Z_o + Z_2 Z_o \right) \left(Z_1 Z_2 + Z_1 Z_0 + 2Z_2 Z_o + Z_o^2 \right)} \qquad (2.107)$$

$$S_{21} = \frac{2Z_2 Z_0}{Z_1 Z_2 + Z_1 Z_o + 2Z_2 Z_0 + Z_o^2} \qquad (2.108)$$

Note that $S_{21} = S_{12}$ for passive networks but that $S_{22} \neq S_{11}$. The output scattering parameter, S_{22}, is therefore:

$$S_{22} = \frac{Z_1 Z_2 - Z_1 Z_o - Z_o^2}{Z_1 Z_2 + Z_1 Z_o + 2Z_2 Z_o + Z_o^2} \qquad (2.109)$$

2.7.2   Simpler Method for Calculating S Parameters

There is a much simpler way of calculating S_{21} or S_{12} by producing a simple equivalent circuit model for the source voltage and source impedance. Remember that for $Z_{01} = Z_{02}$:

$$S_{21} = \frac{V_{out}}{Vi_1} = \frac{V_{out}}{Vi_1} \cdot \frac{V_{in}}{V_{in}} = \frac{V_{out}}{Vi_1} \cdot \frac{(Vi_1 + Vr_1)}{Vi_1 + Vr_1} \qquad (2.110)$$

$$\frac{V_{out}}{V_{in}} \left(1 + \frac{Vr_1}{Vi_1} \right) = \frac{V_{out}}{V_{in}} \left(1 + S_{11} \right) \qquad (2.111)$$

The source can be modelled as a voltage source with a source impedance Z_0 as shown in Figure 2.12.

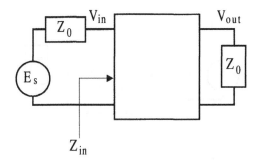

Figure 2.12 Initial model for S parameter calculations

V_{in} can be calculated in terms of Z_0, Z_{in} and E_s as:

$$V_{in} = \frac{Z_{in}}{Z_o + Z_{in}} E_S \qquad (2.112)$$

$$S_{21} = \frac{V_{out}}{V_{in}}\left(1 + S_{11}\right) = V_{out} \, x \, \frac{Z_o + Z_{in}}{Z_{in} E_S}\left(1 + \frac{Z_{in} - Z_o}{Z_{in} + Z_o}\right) \qquad (2.113)$$

$$= \frac{V_{out}}{Z_{in}} \frac{Z_o + Z_{in}}{E_S}\left(\frac{Z_{in} + Z_o}{Z_{in} + Z_o} + \frac{Z_{in} - Z_o}{Z_{in} + Z_o}\right) \qquad (2.114)$$

Therefore:

$$S_{21} = \frac{2V_{out}}{E_S} \qquad (2.115)$$

By making $E_s = 2$ then $S_{21} = V_{out}$:

So a model for calculating the two port S parameters consists of a 2 volt source in series with Z_0 and S_{21} now becomes V_{out} as shown in Figure 2.13.

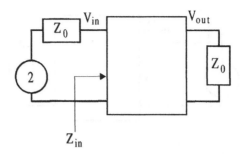

Figure 2.13 Final model for S parameter calculations

The input reflection coefficient can also be calculated in terms of V_{in}, however, it can be shown that this does not save any time in the calculation as it is just as complicated.

$$S_{11} = \frac{Z_{in} - Z_o}{Z_{in} + Z_o} \tag{2.116}$$

$$V_{in} = \frac{Z_{in} - Z_o}{Z_{in} + Z_0} + \frac{Z_{in} + Z_o}{Z_{in} + Z_o} \tag{2.117}$$

$$V_{in} = S_{11} + 1 \tag{2.118}$$

$$S_{11} = V_{in} - 1 \tag{2.119}$$

Now recalculate the S parameters for the same example using the new model as shown in Figure 2.14. Remember that this assumes that $Z_{01} = Z_{02}$.

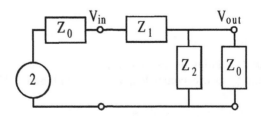

Figure 2.14 Simpler calculation of S parameters

$$S_{21} = V_{out} = 2 \times \frac{\dfrac{Z_o Z_2}{Z_o + Z_2}}{Z_o + Z_1 + \dfrac{Z_o Z_2}{Z_o + Z_2}} = \frac{2 Z_o Z_2}{2 Z_o Z_2 + Z_1 Z_2 + Z_1 Z_o + Z_o^{\,2}} \quad (2.120)$$

which is much simpler than the previous calculation. The input reflection coefficient is:

$$S_{11} = \rho = \frac{Z_{in} - Z_o}{Z_{in} + Z_o} = (V - 1) \quad (2.121)$$

This still requires calculating $V_{in} = 2 Z_{in}/(Z_{in} + Z_0)$ and subtracting one, which saves no time.

2.7.3   S Parameter Summary

1.   The device does not have to be matched to the load to obtain no reverse wave, it should just be terminated in the characteristic impedance of the measurement.

2.   For the same terminating impedance on the input and output = Z_0, S_{21} is not V_{out}/V_{in}; it is V_{out} for a source voltage of 2 volts.

3.   It is also:

$$S_{21} = \frac{V_{out}}{V_{out} \text{ when the device is not there}}$$

This occurs because, when the device is not there, the 2 volt source in series with Z_0 is terminated in Z_0 thereby setting the input voltage to 1 volt.

4.   The forward scatter parameter in terms of power is:

$$\left| S_{21} \right|^2 = \frac{P_L}{P_{avs}}$$

where P_L is the power delivered to the load and P_{avs} is the power available from the source.

2.8   Attenuators (Pads)

It is often important to adjust the level of signals within different parts of a system while still maintaining the correct impedance. For this attenuators are used. These provide a specified attenuation while ensuring that the input and output of the network remains Z_0 as long as the other port is also terminated in Z_0.

Attenuators also offer isolation between circuits and can therefore be used to improve the 'match' between circuits ensuring low reflected power. A measure of match is called the return loss which is $10\log|S_{11}|^2$. This is just the ratio of reflected power from a load over the incident power and is often quoted with the VSWR. Typically a return loss of better than 20dB is required which means that the reverse power is less than 1/100th of the incident power. (The reverse voltage wave is of course 1/10th of the incident wave.) A 20dB return loss is equivalent to a VSWR of 1.22.

T and π versions are shown in Figure 2.15 and 2.16 respectively. The attenuation $= |S_{21}|^2$. This is in fact the transducer gain in this instance.

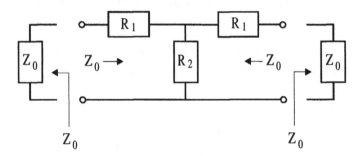

Figure 2.15 Resistive T attenuator

The values of the resistors for a required S_{21} are:

$$R_1 = Z_0 \frac{1 - S_{21}}{1 + S_{21}} \tag{2.122}$$

$$R_2 = Z_0 \frac{2S_{21}}{1 - S_{21}^{\ 2}} \tag{2.123}$$

The π attenuator circuit is shown in Figure 2.16.

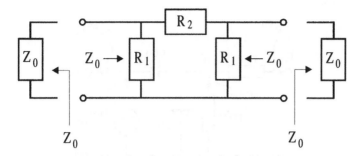

Figure 2.16 Resistive pi attenuator

The values of the resistors for a required S_{21} are:

$$R_1 = Z_0 \frac{1 + S_{21}}{1 - S_{21}}$$ (2.124)

$$R_2 = Z_0 \frac{1 - S_{21}{}^2}{2 S_{21}}$$ (2.125)

2.9   Questions

1.   Calculate the y, z, h and S parameters for the following circuits:

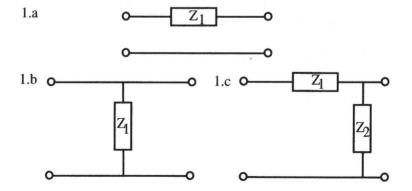

2.   The hybrid π model of a bipolar transistor at RF frequencies is shown
        below:

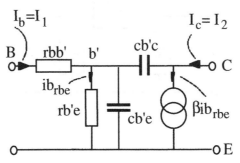

        Derive equations for the short circuit current gain (h_{21}) showing the
        variation with frequency. State any assumptions that you make! Sketch
        this result versus frequency describing all parts of the graph and stating
        the slope of the parts of the curve. What is h_{fe} and calculate f_β (the 3dB
        break point) and f_T. How is f_T measured?

3.   Assuming that a power FET can be represented by the following
        structure, prove that:

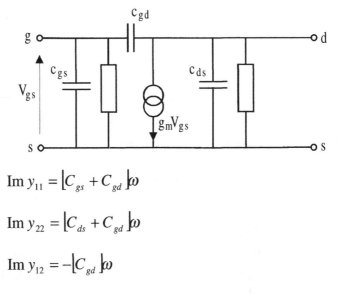

$$\text{Im } y_{11} = [C_{gs} + C_{gd}]\omega$$

$$\text{Im } y_{22} = [C_{ds} + C_{gd}]\omega$$

$$\text{Im } y_{12} = -[C_{gd}]\omega$$

        where Im refers to the imaginary part.

4. A bipolar transistor model has the following y parameters:

$y_{11} = j\omega.1.10^{-12} + 0.01$
$y_{22} = 0.001$
$y_{12} = 0$
$y_{21} = 0.1$

The admittance parameters are in Siemens.

Draw an equivalent circuit for this model (using capacitors, resistors and a controlled current source). What is unusual about this device?

5. Calculate the resistor values R_1 and R_2 for the T and π attenuators in terms of S_{21} and Z_0. Hint: The calculation for the T network can be greatly simplified by writing the input impedance to the network only in terms of Z_0 and R_1. A similar technique can be applied to the π network.

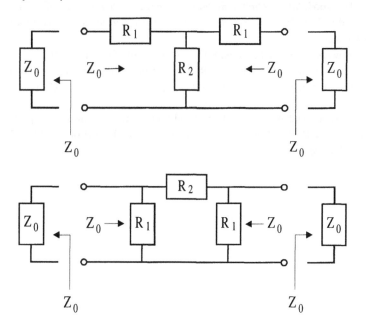

6.   Calculate the S parameters for:

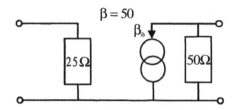

What is the likely collector current through this device if it is assumed to be an ideal bipolar transistor?

2.10   Bibliography

1. R.S. Carson, *High Frequency Amplifiers*, Wiley Interscience, 1982.
2. Guillermo Gonzalez, *Microwave Transistor Amplifiers, Analysis and Design*, Prentice Hall, 1984.
3. W.H. Haywood, *Introduction to Radio Frequency Design*, Prentice Hall, 1982.
4. Hayward and DeMaw, *Solid State Design for the Radio Amateur*, US Amateur Radio Relay League.
5. Chris Bowick, *RF Circuit Design*, SAMS, Division of Macmillan, 1987.
6. H.L. Krauss, C.W. Bostian and F.H. Raab, *Solid State Radio Engineering*, Wiley, 1980.
7. Peter C.L. Yip, *High Frequency Circuit Design and Measurements*, Chapman and Hall, 1990.
8. J.K.A. Everard and A.J. King, "Broadband Power Efficient Class E Amplifiers with a Non-linear CAD Model of the Active MOS Devices", Journal of the IERE, **57**, No. 2, pp. 52–58, 1987.

3

Small Signal Amplifier Design and Measurement

3.1   Introduction

So far device models and the parameter sets have been presented. It is now important to develop the major building blocks of modern RF circuits and this chapter will cover amplifier design. The amplifier is usually required to provide low noise gain with low distortion at both small and large signal levels. It should also be stable, i.e. not generate unwanted spurious signals, and the performance should remain constant with time.

A further requirement is that the amplifier should provide good reverse isolation to prevent, for example, LO breakthrough from re-radiating via the aerial. The input and output match are also important when, for example, filters are used as these require accurate terminations to offer the correct performance. If the amplifier is being connected directly to the aerial it may be minimum noise that is required and therefore the match may not be so critical. It is usually the case that minimum noise and optimum match do not occur at the same point and a circuit technique for achieving low noise and optimum match simultaneously will be described.

For an amplifier we therefore require:

1.   Maximum/specified gain through correct matching and feedback.

2.   Low noise.

3.   Low distortion.

4.   Stable operation.

5.   Filtering of unwanted signals.

6.   Time independent operation through accurate and stable biasing which
        takes into account device to device variation and drift effects caused by
        variations in temperature and ageing.

It has been mentioned that parameter manipulation is a great aid to circuit
design and in this chapter we will concentrate on the use of y and S parameters for
amplifier design. Both will therefore be described.

3.2   Amplifier Design Using Admittance Parameters

A y parameter representation of a two port network is shown in Figure 3.1. Using
these parameters, the input and output impedances/admittances can be calculated
in terms of the y parameters and arbitrary source and load admittances. Stability,
gain, matching and noise performance will then be discussed.

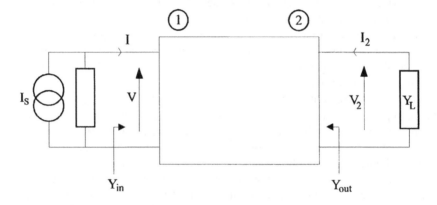

Figure 3.1 y parameter representation of an amplifier

The basic y parameter equations for a two port network are:

$$I_1 = y_{11} V_1 + y_{12} V_2 \qquad\qquad (3.1)$$

$$I_2 = y_{22} V_2 + y_{21} V_1 \qquad\qquad (3.2)$$

From equation (3.1):

$$Y_{in} = \frac{I_1}{V_1} = y_{11} + \frac{y_{12}V_2}{V_1} \tag{3.3}$$

Calculate V_1 from equation 3.2:

$$V_1 = \frac{I_2 - y_{22}V_2}{y_{21}} \tag{3.4}$$

Substituting (3.4) in (3.3):

$$Y_{in} = \frac{I_1}{V_1} = y_{11} + \frac{y_{12}V_2\,y_{21}}{I_2 - y_{22}V_2} \tag{3.5}$$

Dividing top and bottom by V_2:

$$Y_{in} = \frac{I_1}{V_1} = y_{11} + \frac{y_{12}\,y_{21}}{-y_L - y_{22}} = y_{11} - \frac{y_{12}\,y_{21}}{y_L + y_{22}} \tag{3.6}$$

Similarly for Y_{out}:

$$Y_{out} = y_{22} - \frac{y_{21}\,y_{12}}{Y_S + y_{11}} \tag{3.7}$$

Y_{in} can therefore be seen to be dependent on the load admittance Y_L. Similarly Y_{out} is dependent on the source admittance Y_S. The effect is reduced if y_{12} (the reverse transfer admittance) is low. If y_{12} is zero, Y_{in} becomes equal to y_{11} and Y_{out} becomes equal to y_{22}. This is called the unilateral assumption.

3.2.1   Stability

When the real part $Re(Y_{in})$ and/or $Re(Y_{out})$ are negative the device is producing a negative resistance and is therefore likely to be unstable causing potential oscillation. If equations (3.6) and (3.7) are examined it can be seen that any of the parameters could cause instability. However, if y_{11} is large, this part of the input impedance is lower and the device is more likely to be stable. In fact placing a resistor across (or sometimes in series with) the input or output or both is a common method to ensure stability. This degrades the noise performance and it is often preferable to place a resistor only across the output. Note that as y_{12} tends to zero this also helps as long as the real part of y_{11} is positive. The device is unconditionally stable if for all positive g_s and g_L the real part of Y_{in} is greater than zero and the real part of Y_{out} is greater than zero. The imaginary part can of course be positive or negative. In other words the real input and output impedance is always positive for all source and loads which are not negative resistances. Note that when an amplifier is designed the stability should be checked at all frequencies as the impedance of the matching network changes with frequency.

An example of a simple stability calculation showing the value of resistor required for stability is shown in the equivalent section on S parameters later in this chapter.

John Linvill [13] from Stanford developed the Linvill stability parameter:

$$C = \frac{|y_{12}\,y_{21}|}{2\,g_{11}\,g_{22} - \mathrm{Re}(y_{12}\,y_{21})}$$ (3.8)

where g_{11} is the real part of y_{11}. The device is unconditionally stable if C is positive and less than one. Stern [14] developed another parameter:

$$k = \frac{2(g_{11} + G_S)(g_{22} + G_L)}{|y_{12}\,y_{21}| + \mathrm{Re}(y_{12}\,y_{21})}$$ (3.9)

which is stable if $k > 1$. This is different from the Linvill [13] factor in that the Stern [14] factor includes source and load admittances. The Stern factor is less stringent as it only guarantees stability for the specified loads. Care needs to be taken when using the stability factors in software packages as a large K is sometimes used to define the inverse of the Linvill or Stern criteria.

3.2.1.1 Summary for Stability

To maintain stability the $\mathrm{Re}(Y_{in}) \geq 0$ and the $\mathrm{Re}(Y_{out}) \geq 0$ for all the loads presented to the amplifier over the whole frequency range.

The device is unconditionally stable when the above applies for all $\mathrm{Re}(Y_L)$ ≥ 0 and all $\mathrm{Re}(Y_S) \geq 0$. Note that the imaginary part of the source and load can be any value.

3.2.2   Amplifier Gain

Now examine the gain of the amplifier. The gain is dependent on the internal gain of the device and the closeness of the match that the device presents to the source and load. As long as the device is stable maximum gain is obtained for best match. It is therefore important to define the gain. There are a number of gain definitions which include the 'available power gain' and 'transducer gain'. The most commonly used gain is the transducer gain and this is defined here:

$$G_T = \frac{P_L}{P_{AVS}} = \frac{\text{Power delivered to the load}}{\text{Power available from the source}} \qquad (3.10)$$

To calculate this, the output voltage is required in terms of the input current. Using the block diagram in Figure 3.1.

$$V_1 = \frac{I_S}{Y_S + Y_{in}} = \frac{I_S}{Y_S + y_{11} - \dfrac{y_{12}y_{21}}{Y_L + y_{22}}} \qquad (3.11)$$

$$V_1 = \frac{I_S(Y_L + y_{22})}{(Y_S + y_{11})(Y_L + y_{22}) - y_{12}y_{21}} \qquad (3.12)$$

To calculate V_2 remember that:

$$I_2 = y_{22}V_2 + y_{21}V_1 \qquad (3.2)$$

Taking (3.2) therefore:

$$V_2 = \frac{I_2 - y_{21}V_1}{y_{22}} \qquad (3.13)$$

As V_2 is also equal to $-I_2/Y_L$ then $I_2 = -V_2Y_L$ and:

$$V_2 = \frac{-V_2Y_L - y_{21}V_1}{y_{22}} \qquad (3.14)$$

$$V_2 \left(1 + \frac{Y_L}{y_{22}} \right) = \frac{-y_{21}V_1}{y_{22}}$$ (3.15)

$$V_2 = -\left(\frac{y_{21}V_1}{y_{22}} \right)\left(\frac{1}{1 + \frac{Y_L}{y_{22}}} \right)$$ (3.16)

$$V_2 = \frac{-y_{21}V_2}{y_{22} + Y_L}$$ (3.17)

Substituting equation (3.12) in equation (3.17):

$$V_2 = \frac{-I_S\, y_{21}}{(Y_S + y_{11})(Y_L + y_{22}) - y_{12}\, y_{21}}$$ (3.18)

As $P_L = |V_2|^2\, G_L$ where G_L is the real part of Y_L:

$$P_L = \frac{I_S^{\,2} G_L \left| y_{21} \right|^2}{\left|(y_s + y_{11})(Y_L + y_{22}) - y_{12} y_{21}\right|^2}$$ (3.19)

The power available from the source is the power available when matched so:

$$P_{AVS} = \left(\frac{I_S}{2} \right)^2 \frac{1}{G_S}$$ (3.20)

Therefore the transducer gain is:

$$G_T = \frac{P_L}{P_{AVS}} = \frac{4 G_S G_L \left| y_{21} \right|^2}{\left|(Y_S + y_{11})(Y_L + y_{22}) - y_{12} y_{21}\right|^2}$$ (3.21)

For maximum gain we require a match at the input and the output; therefore $Y_s =$ Y_{in}^* and $Y_L = Y_{out}^*$, where * is the complex conjugate.

Remember, however, that as the load is changed so is the input impedance. With considerable manipulation it is possible to demonstrate full conjugate matching on both the input and output as long as the device is stable. The source and load admittances for perfect match are therefore as given in Gonzalez [1]:

$$G_S = \frac{1}{2g_{22}}\left[(2g_{11}g_{22} - \mathrm{Re}(y_{12}y_{21})) - |y_{12}y_{21}|\right]^{\frac{1}{2}} \qquad (3.22)$$

$$B_S = -b_{11} + \frac{\mathrm{Im}(y_{12}y_{21})}{2g_{22}} \qquad (3.23)$$

$$G_L = G_S \frac{g_{22}}{g_{11}} \qquad (3.24)$$

$$B_L = -b_{22} + \frac{\mathrm{Im}(y_{12}y_{21})}{2g_{11}} \qquad (3.25)$$

$$Y_S = G_S + jB_S \qquad Y_L = G_L + jB_L \qquad (3.26)$$

The actual transducer gain for full match requires substitution of equations (3.22) to (3.26) in the G_T equation (3.21).

3.2.3   Unilateral Assumption

A common assumption to ease analysis is to assume that $y_{12}' = 0$, i.e. assume that the device has zero feedback. This is the unilateral assumption where $Y_S = y_{11}^*$ and $Y_L = y_{22}^*$. As:

$$G_T = \frac{P_L}{P_{AVS}} = \frac{4G_S G_L |y_{21}|^2}{|(Y_S + y_{11})(Y_L + y_{22}) - y_{12}y_{21}|^2} \qquad (3.27)$$

$$G_T = \frac{|y_{21}|^2}{4g_{11}g_{22}}$$ (3.28)

This is the maximum unilateral gain often defined as GUM or MUG and is another figure of merit of use in amplifier design. This enables fairly easy calculation of the gain achievable from an amplifier as long as y_{12} is small and this approximation is regularly used during amplifier design.

3.3   Tapped *LC* Matching Circuits

Using the information obtained so far it is now possible to design the matching circuits to obtain maximum gain from an amplifier. A number of matching circuits using tapped parallel resonant circuits are shown in Figure 3.2. The aim of these matching circuits is to transform the source and load impedances to the input and output impedances and all of the circuits presented here use reactive components to achieve this. The circuits presented here use inductors and capacitors.

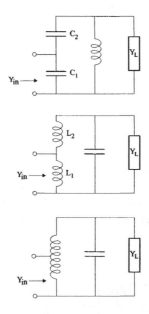

Figure 3.2 Tapped parallel resonant RF matching circuits

A tuned amplifier matching network using a tapped C matching circuits will be presented. This is effectively a capacitively tapped parallel resonant circuit. Both tapped C and tapped L can be used and operate in similar ways. These circuits have the capability to transform the impedance up to the maximum loss resistance of the parallel tuned circuit. The effect of losses will be discussed later.

Two component reactive matching circuits, in the form of an L network, will be described in the section on amplifier design using S parameters and the Smith Chart.

A tapped C matching circuit is shown in Figure 3.2a. The aim is to design the component values to produce the required input impedance, e.g. 50Ω for the input impedance of the device which can be any impedance above 50Ω. To analyse the tapped C circuit it is easier to look at the circuit from the high impedance point as shown in Figure 3.3.

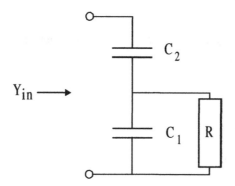

Figure 3.3 Tapped C circuit for analysis

The imaginary part is then cancelled using the inductor. Often a tunable capacitor is placed in parallel with the inductor to aid tuning. Y_{in} is therefore required:

$$Y_{in} = \text{Real} + \text{Imaginary parts} = G + jB \qquad (3.29)$$

Initially we calculate Z_{in}:

$$Z_{in} = \frac{R / sC_1}{R + \dfrac{1}{sC_1}} + \frac{1}{sC_2} \qquad (3.30)$$

and with algebra:

$$Y_{in} = \frac{s^2 C_1 C_2 R + s C_2}{s C_2 R + s C_1 R + 1} \tag{3.31}$$

The real part of Y_{in} is therefore:

$$Y_{in} = \frac{\omega^2 C_2^{\,2} R}{1 + \omega^2 R^2 (C_1 + C_2)} \tag{3.32}$$

The shunt resistive part of Y_{in} is therefore R_{in}:

$$R_{in} = \frac{1 + \omega^2 R^2 (C_1 + C_2)^2}{\omega^2 C_2^{\,2} R} \tag{3.33}$$

If we assume (or ensure) that $\omega^2 R^2 (C_1 + C_2)^2 > 1$, which occurs for loaded Qs greater than 10, then:

$$R_{in} = R \left(1 + \frac{C_1}{C_2} \right)^2 \tag{3.34}$$

The imaginary part of Y_{in} is:

$$Y_{in(imag)} = \frac{s \omega^2 C_1 C_2 R^2 (C_1 + C_2) + s C_2}{1 + \omega^2 R^2 (C_1 + C_2)^2} \tag{3.35}$$

Making the same assumption as above and assuming that C_2 is smaller than $\omega^2 C_1 C_2 R^2 (C_1 + C_2)$ then:

$$C_T = \frac{C_1 C_2}{C_1 + C_2} \tag{3.36}$$

This is equivalent to the two capacitors being added in series.

In conclusion the two important equations are:

$$R_{in} = R\left(1 + \frac{C_1}{C_2}\right)^2 \tag{3.34}$$

and

$$C_T = \frac{C_1 C_2}{C_1 + C_2} \tag{3.36}$$

These can be further simplified:

$$\frac{C_1}{C_2} = \left(\sqrt{\frac{R_{in}}{R}} - 1\right) \tag{3.37}$$

Therefore:

$$C_1 = C_2\left(\sqrt{\frac{R_{in}}{R}} - 1\right) \tag{3.38}$$

As:

$$C_T = \frac{\left(\sqrt{\dfrac{R_{in}}{R}} - 1\right)C_2 C_2}{\left(\sqrt{\dfrac{R_{in}}{R}} - 1\right)C_2 + C_2} \tag{3.39}$$

dividing through by C_2 gives:

$$C_T = \frac{\left(\sqrt{\dfrac{R_{in}}{R}} - 1\right) C_2}{\left(\sqrt{\dfrac{R_{in}}{R}} - 1\right) + 1} \qquad (3.40)$$

Therefore:

$$C_T = \frac{\left(\sqrt{\dfrac{R_{in}}{R}} - 1\right) C_2}{\left(\sqrt{\dfrac{R_{in}}{R}}\right)} \qquad (3.41)$$

and:

$$C_2 = \frac{\left(\sqrt{\dfrac{R_{in}}{R}}\right)}{\left(\sqrt{\dfrac{R_{in}}{R}} - 1\right)} C_T \qquad (3.42)$$

as:

$$C_1 = C_2 \left(\sqrt{\dfrac{R_{in}}{R}} - 1\right) \qquad (3.38)$$

$$C_1 = \left(\sqrt{\dfrac{R_{in}}{R}}\right) C_T \qquad (3.43)$$

3.3.1   Tapped C Design Example

Let us match a 50Ω source to a 5K resistor in parallel with 2pF at 100MHz. A block diagram is shown in Figure 3.4. A 3dB bandwidth of 5 MHz is required. This is typical of the older dual gate MOSFET. This is an integrated four terminal device which consists of a Cascode of two MOSFETS. A special feature of Cascodes is the low feedback C when gate 2 is decoupled. C feedback for most dual gate MOSFETs is around 20 to 25fF. An extra feature is that varying the DC bias on gate 2 varies the gain experienced by signals on gate 1 by up to 50dB. This can be used for AGC and mixing.

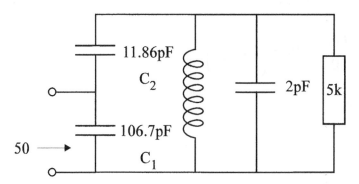

Figure 3.4 Tapped C design example

To obtain the 3dB bandwidth the loaded Q, Q_L is required:

$$Q_L = \frac{R_{total}}{\omega L} = \frac{f_0}{3dB\ BW} = \frac{100}{5} \qquad (3.44)$$

R_{total} is the total resistance across L. This includes the transformed up source impedance in parallel with the input impedance which for a match is equal to 5K/2.

$$L = \frac{R_{total}}{\omega Q_L} = \frac{2k5}{20.2.\pi.10^8} = 200nH \qquad (3.45)$$

Therefore to obtain C_T

$$f = \frac{1}{2\pi\sqrt{LC}} \qquad (3.46)$$

so:

$$LC = \frac{1}{(2\pi f)^2}$$

(3.47)

As $L = 200\text{nH}$ at 100MHz

$$C_{res} = 12.67\text{pF}$$

(3.48)

$$C_T = C_{res} - 2\text{pf} = 10.67\text{pF}$$

(3.49)

$$\frac{R_{in}}{R} = \frac{5000}{50} = 100 = \left(1 + \frac{C_1}{C_2}\right)$$

(3.50)

Therefore:

$$\frac{C_1}{C_2} = 9$$

(3.51)

Thus:

$$C_1 = 9C_2$$

(3.52)

and:

$$C_T = \frac{C_1 C_2}{C_1 + C_2} = \frac{9C_2 C_2}{9C_2 + C_2}$$

(3.53)

$$C_T = \frac{9}{10}C_2$$

(3.54)

$$C_2 = \frac{10}{9}C_T$$

(3.55)

$$C_2 = 11.86\text{pf}$$

(3.56)

$$C_1 = 9C_2 \text{ (or } 10C_T) = 106.7\text{Pf}$$

(3.57)

The approximations can be checked to confirm the correct use of the equations if the loaded Q is less than 10. $\omega^2 R^2 (C_1 + C_2)^2$ should be much greater than one for the approximations to hold. Also ensure that $C_2 \ll \omega^2 C_1 C_2 R^2 (C_1 + C_2)$ for the approximations to hold.

3.4   Selectivity and Insertion Loss of the Matching Network

It is important to consider the effect of component losses on the performance of the circuit. This is because the highest selectivity can only be achieved by making the loaded Q approach the unloaded Q. However, as shown below, the insertion loss tends towards infinity as the loaded Q tends towards the unloaded Q. This is most easily illustrated by looking at a series resonant circuit as shown in Figure 3.5. This consists of an LCR circuit driven by a source and load of Z_0. The resistor in series with the LC circuit is used to model losses in the inductor/capacitor.

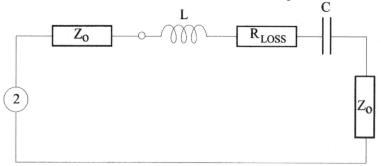

Figure 3.5 *LCR* model for loss in resonant circuits

Using the S parameters to calculate the transducer gain (remember that $S_{21} = V_{out}$ if the source is 2 volts and the source and load impedances are both the same):

$$S_{21} = V_{out} = \frac{2Z_0}{2Z_0 + R_{LOSS} + j\left(\omega L - \dfrac{1}{\omega C}\right)} \qquad (3.58)$$

At resonance:

$$S_{21} = \frac{2Z_0}{2Z_0 + R_{LOSS}} \qquad (3.59)$$

As:

$$Q_0 = \frac{\omega L}{R_{LOSS}}$$ (3.60)

$$R_{LOSS} = \frac{\omega L}{Q_0}$$ (3.61)

As:

$$Q_L = \frac{\omega L}{R_{LOSS} + 2Z_0}$$ (3.62)

$$R_{LOSS} + 2Z_0 = \frac{\omega L}{Q_L}$$ (3.63)

Therefore at resonance:

$$S_{21} = \omega L \frac{\left[\dfrac{1}{Q_L} - \dfrac{1}{Q_0} \right] Q_L}{\omega L}$$ (3.64)

giving:

$$S_{21} = \left(1 - \frac{Q_L}{Q_0} \right)$$ (3.65)

Also note that for $df < f_o$:

$$S_{21df} = \left(1 - \frac{Q_L}{Q_0} \right) \frac{1}{\left(1 \pm 2jQ_L \dfrac{df}{f_o} \right)}$$ (3.66)

This can be used to calculate the frequency response further from the centre frequency. Remember that:

$$G_T = \frac{P_L}{P_{AVS}}$$ (3.67)

$$P_{AVS} = \frac{1^2}{R_S}$$ (3.68)

$$P_L = \frac{(V_{out})^2}{R_L}$$ (3.69)

Therefore as long as $R_L = R_S$:

$$G_T = \frac{P_L}{P_{AVS}} = (V_{out})^2 = |S_{21}|^2$$ (3.70)

As $S_{21} = V_{out}$ for $V = 2$ source voltage:

$$G_T = \left(1 - \frac{Q_L}{Q_0}\right)^2$$ (3.71)

where:

$$Q_L = \frac{\omega L}{R_{TOTAL}} = \frac{\omega L}{R_{LOSS} + 2Z_0}$$ (3.62)

$$Q_0 = \frac{\omega L}{R_{loss}}$$ (3.60)

It is interesting to investigate the effect of insertion loss on this input matching network. For a bandwidth of 5 MHz, $Q_L = 20$. If we assume that $Q_0 = 200$, $G_T = (0.9)^2 = -0.91$dB loss. The variation in insertion loss versus Q_L/Q_0 is shown in Figure 3.6 for four different values of Q_L/Q_0. For finite Q_0, Q_L can be increased towards Q_0 however, the insertion loss (G_T) will tend to infinity.

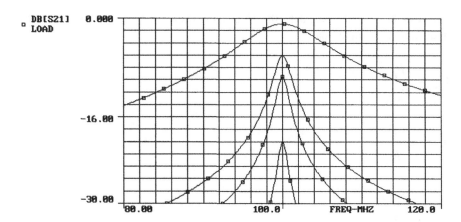

Figure 3.6 Variation in insertion loss for Q_L/Q_0 = (a) 0.1 (top) (b) 0.5 (c) 2/3 (d) 0.9
(bottom)

It is therefore possible to trade selectivity for insertion loss. If low noise is
required the input matching network may be set to a low Q_L to obtain low Q_L/Q_0 as
the insertion loss of the matching circuit will directly add to the noise figure. Note
that for lower transformation ratios this is often not a problem. A plot of S_{21} against
Q_L/Q_0 is shown in Figure 3.7 showing that as the insertion loss tends to infinity S_{21}
tends to zero and Q_L tends to Q_0.

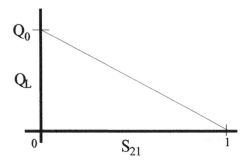

Figure 3.7 S_{21} vs Q_L/Q_0

Measurements of S_{21} vs Q_L offer a way of obtaining Q_0. The intersection on the Y
axis being Q_0. Q_0 for open coils is typically 100 → 300; for open printed coils this
reduces to 20 to 150.

3.5   Dual Gate MOSFET Amplifiers

The tapped C matching circuit can be used for matching dual gate MOSFETs. These are integrated devices which consist of two MOSFETs in cascode. A typical amplifier circuit using a dual gate MOSFET is shown in Figure 3.8. The feedback capacitance is reduced to around 25fF as long as gate 2 is decoupled. Further the bias on gate 2 can be varied to obtain a gain variation of up to 50dB. For an N channel depletion mode FET, 4 to 5 volts bias on gate 2 (V_{G2S}) gives maximum gain.

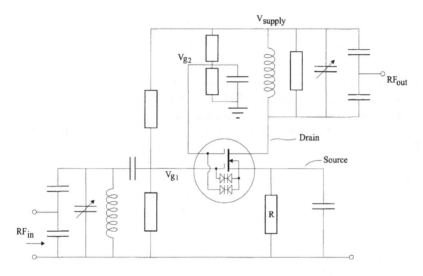

Figure 3.8 Dual gate MOSFET amplifier

As an example it is interesting to investigate the stability of the BF981. Taking the Linvill [13] stability factor:

$$C = \frac{|y_{12} y_{21}|}{2 g_{11} g_{22} - \mathrm{Re}(y_{12} y_{21})} \tag{3.67}$$

where the device is unconditionally stable when C is positive and less than one. We apply this to the device at 100MHz using the y parameters from the data sheets:

$$y_{21} = 20 \times 10^{-3} \angle 6° = (19.89 + 2j) 10^{-3} \tag{3.68}$$

$$y_{12} = 13.10^{-6} \angle 90° = 13.10^{-6} j = 20\text{fF} \tag{3.69}$$

$$g_{11} = 45 \times 10^{-6} \approx 22\text{k}\Omega \tag{3.70}$$

$$g_{22} = 45 \times 10^{-6} \approx 22\text{k}\Omega \tag{3.71}$$

Therefore the Linvill [13] stability factor predicts:

$$C = \frac{\left|13j \times (19.89 + 2j)\right| \times 10^{-9}}{\left(4.05 \times 10^{-9}\right) - \left(-26 \times 10^{-9}\right)} = \frac{260 \times 10^{-9}}{30 \times 10^{-9}} \tag{3.72}$$

The device is therefore not unconditionally stable as C is greater than one. This is because the feedback capacitance (20fF) although low, still presents an impedance of similar value to the input and output impedances.

To ensure stability it is necessary to increase the input and output admittances effectively by lowering the resistance across the input and output. This is achieved by designing the matching network to present a much lower resistance across the input and output. Shunt resistors can also be used but these degrade the noise performance if used at the input. Therefore we look at Stern [14] stability factor which includes source and load impedances, where stability occurs for $k > 1$.

$$k = \frac{2(g_{11} + G_S)(g_{22} + G_L)}{\left|y_{12} y_{21}\right| + \text{Re}(y_{12} y_{21})} \tag{3.73}$$

$$\left|y_{12} y_{21}\right| = 260 \times 10^{-9} \tag{3.74}$$

$$\text{Re}(y_{12} y_{21}) = -26 \times 10^{-9} \tag{3.75}$$

As the device is stable for $k > 1$ it is possible to ensure stability by making $2(g_{11} + G_S)$ $(g_{22} + G_L) > 234 \times 10^{-9}$. One method to ensure stability is to place equal admittances on the I/P and O/P. To achieve this the total input admittance and output admittance are each 3.4×10^{-4} i.e. 2.9kΩ. This of course just places the device on the border of stability and therefore lower values should be used. The source and load impedances could therefore be transformed up from, say, 50Ω to 2kΩ. The match will also be poor unless resistors are also placed across the input and output of the device. The maximum available gain is also reduced but this is usually not a problem as the intrinsic matched gain is very high at these frequencies. It is also necessary to calculate the stability factors at all other frequencies as the impedances presented across the device by the matching networks will vary considerably with frequency. It will be shown in the next section that the noise performance is also dependent on the source impedance and in fact for this device the real part of the optimum source impedance for minimum noise is 2kΩ.

3.6   Noise

The major noise sources in a transistor are:
1.   Thermal noise caused by the random movement of charges.
2.   Shot noise.
3.   Flicker noise.

The noise generated in an amplifier is quantified in a number of ways. The noise factor and the noise figure. Both parameters describe the same effect where the noise figure is 10 log(noise factor). This shows the degradation caused by the amplifier where an ideal amplifier has a noise factor of 1 and a noise figure of 0dB. The noise factor is defined as:

$$NF = \frac{\text{Total available output noise power}}{\text{Available noise output arising from the thermal noise in the source}} = \frac{P_{no}}{G_A P_{ni}} \quad (3.76)$$

Where G_A is the available power gain and P_{ni} is the noise available from the source. The noise power available from a resistor at temperature T is kTB, where k is Boltzmann constant, T is the temperature and B is the bandwidth. From this the equivalent noise voltage or noise current for a resistor can be derived. Let us assume that the input impedance consists of a noiseless resistor driven by a conventional resistor. The conventional resistor can then be represented either as a noiseless resistor in parallel with a noise current or as a noiseless resistor in series with a noise voltage as shown in Figures 3.9 and 3.10 where:

$$i_n^2 = \frac{4kTB}{R} \quad (3.77)$$

and:

$$e_n^2 = 4kTBR \quad (3.78)$$

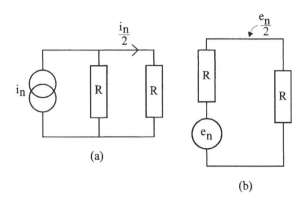

(a)

(b)

Figure 3.9 Equivalent current noise source. **Figure 3.10** Equivalent voltage noise source

Note that there is often confusion about the noise developed in the input impedance of an active device. This is because this impedance is a dynamic AC impedance not a conventional resistor. In other words, r_e was dependent on dV/dI rather than V/I. For example, if you were to assume that the input impedance was made up of 'standard resistance' then the minimum achievable noise figure would be 3dB. In fact the noise in bipolar transistors is caused largely by 'conventional' resistors such as the base spreading resistance $r_{bb'}$, the emitter contact resistance and shot noise components.

In active devices the noise can most easily be described by referring all the noise sources within the device back to the input. A noisy two port device is often modelled as a noiseless two port device with all the noises within the device transformed to the input as a series noise voltage and a shunt noise current as shown in Figure 3.11.

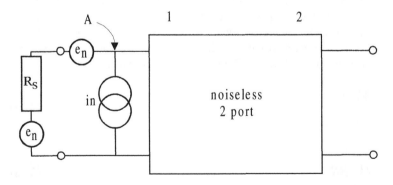

Figure 3.11 Representation of noise in a two port

It is now worth calculating the optimum source resistance, R_{SO}, for minimum noise figure. The noise factor for the input circuit is obtained by calculating the ratio of the total noise at node A to the noise caused only by the source impedance R_S:

$$NF = \frac{4kTBR_S + e_n^{\ 2} + (i_n R_S)^2}{4kTBR_S} \tag{3.79}$$

Therefore:

$$NF = 1 + \frac{e_n^{\ 2} + (i_n R_S)^2}{4kTBR_S} \tag{3.80}$$

Differentiating the noise factor with respect to R_S:

$$\frac{dNF}{dR_S} = \frac{1}{4kTB}\left(\frac{-e_n^{\ 2}}{R_S^{\ 2}} + i_n^{\ 2}\right) \tag{3.81}$$

Equating this to zero means that:

$$\frac{e_n^{\ 2}}{R_S} = i_n^{\ 2} \tag{3.82}$$

Therefore the optimum source impedance for minimum noise is:

$$R_{SO} = \frac{e_n}{i_n} \tag{3.83}$$

The minimum noise figure, F_{min}, for uncorrelated sources is therefore obtained by substituting equation (3.83) in (3.80).

$$NF = 1 + \frac{e_n^{\ 2} + \left(i_n \dfrac{e_n}{i_n}\right)^2}{4kTB\dfrac{e_n}{i_n}} = 1 + \frac{2e_n^{\ 2} i_n}{4kTBe_n} \tag{3.84}$$

Therefore:

$$F_{min} = 1 + \frac{e_n i_n}{2kTB}$$ (3.85)

There is therefore an optimum source impedance for minimum noise. This effect can be shown to produce a set of noise circles. An example of the noise circles for the BF981 dual gate MOSFET is shown in Figure 3.12 where the optimum source impedance for minimum noise at 100MHz can be seen to be at the centre of the circle where: $G_s = 0.5 \times 10^{-3}$ and $B_s = -1j \times 10^{-3}$. This is equivalent to an optimum source impedance represented as a 2kΩ resistor in parallel with an inductor of 1.6μH (at 100MHz).

Note that these values are far away from the input impedance which in this device can be modelled as a 22kΩ resistor in parallel with 2pf. This illustrates the fact that impedance match and optimum noise match are often at different positions. In fact this effect is unusually exaggerated in dual gate MOSFETs operating in the VHF band due to the high input impedance. For optimum sensitivity it is therefore more important to noise match than to impedance match even though maximum power gain occurs for best impedance match. If the amplifier is to be connected directly to an aerial then optimum noise match is important. In this case that would mean that the aerial impedance should be transformed to present 2K in parallel with 1.6uH at the input of the device which for low loss transformers would produce a noise figure for this device of around 0.6dB. Losses in the transformers would be dependent on the ratio of loaded Q to unloaded Q. Note that the loss resistors presented across the tuned circuit would not now be half the transformed impedance (2k) as impedance match does not occur, but 2kΩ in parallel with 22kΩ.

There is a further important point when considering matching and that is the termination impedance presented to the preceding device. For example if there was a filter between the aerial and the amplifier, the filter would only work correctly when terminated in the design impedances. This is because a filter is a frequency dependent potential divider and changing impedances would change the response and loss.

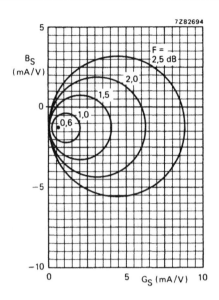

Figure 3.12 Noise circles for the BF981. Reproduced with permission from Philips using data book SC07 on Small Signal Field Effect Transistors

It should be noted that at higher frequencies the noise sources are often partially correlated and then the noise figure is given by [1] and [11]:

$$F = F_{min} + \frac{r_n}{g_s}\left|Y_S - Y_0\right|^2 \tag{3.86}$$

where r_n is the normalised noise resistance:

$$r_n = \frac{R_N}{Z_0} \tag{3.87}$$

Note that the equivalent noise resistances are concept resistors which can be used to represent voltage or current noise sources. This is the value of resistor having a thermal noise equal to the noise of the generator at a defined temperature.

Therefore:

$$R_{ne} = \frac{e_n^2}{4kTB}$$ (3.88)

$$R_{ni} = \frac{4kTB}{i_n^2}$$ (3.89)

$Y_s = g_s + jb_s$ represents the source admittance   (3.90)

$Y_0 = g_0 + jb_0$ represents the source admittance which results in minimum noise figure. These parameters can be converted to reflection coefficients for the source and load admittances:

$$Y_S = \frac{1 - \Gamma_s}{1 + \Gamma_s}$$ (3.91)

$$Y_0 = \frac{1 - \Gamma_0}{1 + \Gamma_0}$$ (3.92)

$$F = F_{min} + \frac{4r_n|\Gamma_S - \Gamma_0|^2}{\left(1 - |\Gamma_S|^2\right)|1 + \Gamma_0|^2}$$ (3.93)

These parameters are often quoted in S parameter files. An example of the typical parameters for a BFG505 bipolar transistor is shown in Table 1. The S parameters versus frequency are shown at the top of the file and the noise parameters at the bottom. The noise parameters are from left to right: frequency, F_{min}, Γ_{opt} in terms of the magnitude and angle and r_n.

Table 3.1 Typical S parameter and noise data for the BFG505 transistor operating at 3Vand 2.5mA. Reproduced with permission from Philips, using the RF wideband transistors product selection 2000, discrete semiconductors CD.

! Filename: BFG505C.S2P   Version: 3.0
! Philips part #: BFG505   Date: Feb 1992
! Bias condition: Vce=3V, Ic=2.5mA
! IN LINE PINNING: same data as with cross emitter pinning.
MHz S MA R 50

! Freq	S11		S21		S12		S22		!GUM [dB]
40	.949	-3.5	6.588	176.1	.005	86.6	.993	-2.0 !	45.1
100	.942	-8.7	6.499	170.9	.012	83.9	.988	-4.9 !	42.0
200	.920	-17.1	6.337	163.0	.024	79.0	.972	-9.5 !	36.8
300	.892	-25.6	6.226	155.7	.035	74.1	.949	-13.8 !	32.8
400	.858	-33.9	6.046	149.3	.045	69.6	.922	-17.6 !	29.6
500	.823	-41.3	5.771	143.5	.053	65.8	.893	-21.1 !	27.1
600	.788	-48.4	5.529	138.3	.060	62.9	.862	-24.1 !	25.0
700	.750	-55.5	5.338	133.3	.066	60.0	.830	-26.5 !	23.2
800	.706	-62.1	5.126	128.1	.071	57.6	.801	-28.6 !	21.6
900	.663	-68.1	4.858	123.3	.076	55.6	.772	-30.5 !	20.2
1000	.619	-74.2	4.605	119.0	.080	53.8	.745	-32.3 !	18.9
1200	.539	-86.9	4.210	111.0	.088	51.6	.702	-35.9 !	16.9
1400	.480	-99.2	3.910	103.9	.094	49.9	.675	-38.8 !	15.6
1600	.436	-109.5	3.550	97.7	.099	49.6	.656	-40.9 !	14.4
1800	.388	-118.1	3.232	93.3	.104	49.8	.633	-42.1 !	13.1
2000	.337	-129.1	2.967	88.3	.107	49.2	.604	-43.8 !	11.9
2200	.307	-142.7	2.770	83.9	.112	48.6	.577	-46.9 !	11.0
2400	.304	-154.6	2.585	78.5	.115	48.6	.566	-51.0 !	10.3
2600	.304	-163.0	2.386	75.5	.123	49.7	.576	-54.8 !	9.7
2800	.288	-170.1	2.291	72.2	.129	50.2	.588	-56.5 !	9.4
3000	.275	179.0	2.125	68.6	.131	51.1	.582	-57.3 !	8.7

! Noise data:

! Freq.	Fmin	Gamma-opt		rn
900	1.30	.583	19.0	.69
2000	1.90	.473	45.0	.55

As mentioned earlier, it is important to note that maximum gain, optimum match and minimum noise very rarely occur at the same point within an active device. For minimum noise, the match is not critical, but if the termination of the input filter is important then simultaneous noise and impedance match are important.

A circuit technique that can be used to improve this situation in bipolar transistors is the addition of an emitter inductor. This is illustrated by looking at the progression of the hybrid π model to a T model which incorporates a complex gain as shown in Figure 3.13. This is similar to the analysis described in Hayward [2] although the initial approximations are different.

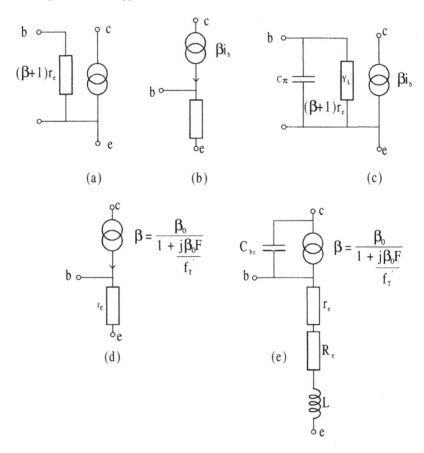

Figure 3.13 Evolution of the Hybrid π model using complex gain

The simple low frequency π model is shown in Figure 3.13.a. and this is directly equivalent to the T model shown in Figure 3.13.b. Now add the input capacitance as shown in Figure 3.13c. This is now equivalent to the model shown in Figure 3.13d. where a complex gain is used to incorporate the effect of the input capacitance. This produces a roll-off which is described by f_T'. f_T' is a modified f_T

because the value of f_T is measured into a S/C and therefore already incorporates the feedback capacitor in parallel with Cπ.

$$f_T' = \frac{C_\pi + C_{bc}}{C_\pi} f_T \tag{3.94}$$

The feedback capacitor can then be added to the T model to produce the complex T model shown in Figure 3.13.e. Note that for ease, the unilateral assumption can be made and the feedback capacitor completely ignored and then f_T' can be assumed to be f_T. This is assumed in the calculations performed here.

Take as an example the fourth generation bipolar transistor BFG505 which has an f_T of 6.5 GHz at 2.5 mA and 3V bias for which Philips provides a design using simulation and measurement. We will show here that a simple analysis can provide accurate results.

Using the S parameter table for the BFG505, the optimum source impedance at 900MHz is $\Gamma opt = 0.583$ angle 19 degrees which is 50(3.2 + 1.5j) = 160Ω + 75jΩ.

Using the simple scalar model, assume the device is unilateral and take the equations for complex β as shown if Figure 3.13; then $Z_{in} = (\beta + 1)Z_e$. If $(\beta + 1)$ is represented as $(A + jB)$ and Z_e is$(r_e + j\omega L)$ then the input impedance is $Z_{in} = (A + jB)(r_e + j\omega L)$. It can therefore be seen that the real part of the input impedance can be increased significantly by the addition of the emitter inductor. Taking equations for β at 900 MHz and assuming that the f_T is 6.5 GHz and β_0 is 120 then $\beta = (0.43 - 7.2j)$, $(\beta + 1) = (1.43 - 7.2j)$. At 2.5 mA $r_e = 25/2.5 = 10\Omega$. $Z_{in} = (\beta + 1)Z_e = (1.43 -7.2j)(10 + j\omega L)$. To obtain a 160$\Omega$ real part then 7.2j.jωL should equal (160 - 14) = 146. This gives a value for L of 3.6nH.

This is very close to the design, simulation and measurement of a 900 MHz amplifier described in the Philips CD on RF wideband transistors entitled 'Product Selection 2000 Discrete Transistors', Application note 10 (SC14), and illustrates how simple models can be used to give a good insight into RF design.

3.6.1 Noise Temperature

Amplifier noise can also be modelled using the concept of noise temperature. The amplifier is modelled as an ideal amplifier with a summing junction at the input. The input to this summing junction consists of the source noise, kT_0B, where T_0 is the ambient temperature of the source, and an equivalent noise source of value kT_eB which represents the degradation caused by the amplifier as shown in Figure 3.14. Note therefore that the noise temperature of a perfect amplifier is zero kelvin.

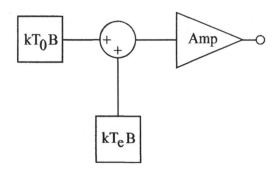

Figure 3.14 Amplifier representation of noise temperature

The noise factor in terms of the noise temperature is:

$$NF = \frac{G_A\left(kT_0B + kT_eB\right)}{kT_0BG_A} = 1 + \frac{T_e}{T_0} \qquad (3.95)$$

It can be seen that the noise temperature is independent of the temperature of the source resistance unlike the noise factor, however, it is dependent on the value of source resistance and the temperature of the amplifier.

3.6.2   Noise Measurement System

Modern noise measurement systems utilise a noise source which can be switched between two discrete values of noise power connected to the input of the device under test (DUT). The output noise power of the DUT is then measured and the change in output noise power measured when the input noise power is switched. If for example an amplifier had zero noise figure and therefore contributed no noise then the change in output power would be the same as the change in input power. If the amplifier had a high noise figure (i.e. it produced a significant amount of excess noise) then the change in output noise power would be much smaller due to the masking effect of the amplifier noise. The system presented here is based on absolute temperature and offers a very simple measurement technique. The system is shown in Figure 3.15.

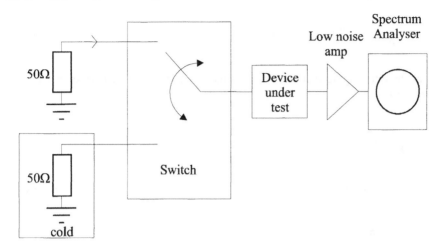

Figure 3.15 Noise Measurement System

The system consists of two 50Ω sources one at room temperature and one placed in liquid nitrogen at 77K. The room temperature and cold resistors are connected sequentially to the amplifier under test and the output of the DUT is applied to a low noise amplifier and spectrum analyser. The change in noise power is measured.

This change in noise power P_R can be used to calculate the noise temperature directly from the following equation by taking the ratio of the sum of the noise sources at the two source temperatures:

$$\frac{T_e + T_{01}}{T_e + T_{02}} = \frac{T_e + 290\text{K}}{T_e + 77\text{K}} = P_R \tag{3.96}$$

T_e is the noise temperature of the amplifier at the operating temperature. T_{01} is the higher temperature of the source in this case room temperature and T_{02} is the temperature of liquid nitrogen. The noise factor can be obtained directly:

$$NF = 1 + \frac{T_e}{T_{01}} = 1 + \frac{T_e}{290\text{K}} \tag{3.97}$$

For an amplifier which contributes no noise (the noise temperature $T_e = 0$, the noise factor is 1, and the noise figure is 0dB) the change in power can be seen to be:

$$P_R = \frac{290}{77} = 3.766 = 5.75\text{dB} \tag{3.98}$$

So as the noise figure increases this change in power is reduced. To calculate T_e in terms of P_R:

$$\frac{T_e + T_{01}}{T_e + T_{02}} = P_R \tag{3.99}$$

$$T_e + T_{01} = P_R\left(T_e + T_{02}\right) \tag{3.100}$$

$$T_e\left(1 - P_R\right) = P_R.T_{02} - T_{01} \tag{3.101}$$

$$T_e = \frac{\left(P_R \times T_{02}\right) - T_{01}}{1 - P_R} = \frac{290 - \left(P_R \times 77\right)}{P_R - 1} \tag{3.102}$$

Plots of P_R vs noise temperature, P_R vs noise factor and P_R vs noise figure are shown in Figures 3.16, 3.17 and 3.18 respectively.

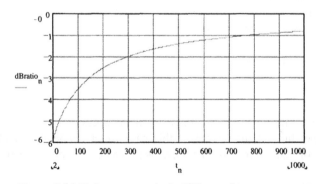

Figure 3.16 Noise power ratio P_R (dB) vs noise temperature

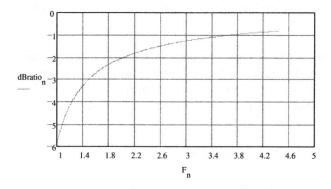

Figure 3.17 Noise power ratio P_R vs noise factor

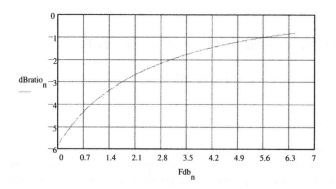

Figure 3.18 Noise power ratio P_R vs noise figure

Take an example of a measured change in output noise power of 3.8dB when the source resistors are switched from cold to hot. Using Figure 3.18, this would predict a noise figure of 1dB.

Note that to obtain accurate measurements the device under test should be mounted in a screened box (possibly with battery power). If the detector consisted of a spectrum analyser then a low noise amplifier would be required at the analyser input as most spectrum analysers have noise figures of 20 to 30dB. The effect of detector noise figure can be deduced from the noise figure of cascaded amplifiers. The losses in the cables connected to the resistors and the switch should be kept low.

3.7   Amplifier Design Using S Parameters and the Smith Chart

For amplifier design at higher frequencies the device characteristics are usually provided using S parameters. Further, most modern measurements, taken above 5 MHz, are made using S parameter network analysers. It is therefore important to understand the equivalent y parameter equations but now using S parameters. This section will cover:

1.   The Smith Chart[1] calculator.

2.   Input and output reflection coefficients/stability and gain.

3.   Matching using Smith Charts.

4.   Broadband Amplifiers.

5.   DC biasing of bipolar transistors and GaAs FETs.

6. S parameter measurements and error correction.

3.7.1   The Smith Chart

It can be seen that the amplifier design techniques shown so far have used parameter sets which deal in voltages and currents. It was also mentioned that most RF measurements use S parameter network analysers which use travelling waves to characterise the amplifiers. These travelling waves enable reasonable terminating impedances to be used as it is easy to manufacture coaxial cable with characteristic impedances around 50 to 75Ω. This also enables easy interconnection and error correction.

It is important to be able to convert easily from impedance or admittance to reflection coefficient and therefore a graphical calculator was developed. To help in this conversion P.H. Smith, while working at Bell Telephone Laboratories, developed a transmission line calculator (Electronic Vol.12, pp.29-31) published in 1939. This is also described in the book by Philip H Smith entitled 'Electronic Applications of the Smith Chart' [12]. This chart consists of a polar/cartesian plot of reflection coefficient onto which is overlaid circles of constant real and constant imaginary impedance. The standard chart is plotted for $|\rho| \leq 1$. The impedance

---

[1] SMITH is a registered trademark of the Analog Instrument Co, Box 950, New Providence, N.J. 07975, USA.

lines form part circles for constant real and imaginary parts. The derivations for the equations are shown here and consist of representing both ρ and impedance, z, in terms of their real and imaginary parts. Using algebraic manipulation it is shown that constant real parts of the impedance form circles on the ρ plot and that the imaginary parts of the impedance form a different set of circles.

Let:

$$\rho = u + jv \tag{3.103}$$

$$z = r + jx \tag{3.104}$$

as:

$$z = \frac{1+\rho}{1-\rho} \tag{3.105}$$

$$r + jx = \frac{1+u+jv}{1-u-jv} = \frac{1-u^2-v^2+2jv}{(1-u)^2+v^2} \tag{3.106}$$

The following equations are labeled (3.107a/b) to (3.113a/b) where (a) is on the LHS and (b) is on the RHS.

Real Part of Z   **Imaginary part of Z**

$$r = \frac{1-u^2-v^2}{(1-u)^2+v^2} \qquad\qquad x = \frac{2v}{(1-u)^2+v^2}$$

$$r - 2ru + ru^2 + rv^2 = 1 - u^2 - v^2 \qquad\qquad x - 2ux + u^2x + v^2x = 2v$$

$$(r+1)u^2 - 2ru + (r+1)v^2 = 1 - r \qquad\qquad xu^2 - 2xu + xv^2 - 2v = -x$$

$$u^2 - \frac{2ru}{r+1} + v^2 = \frac{1-r}{r+1} \qquad\qquad u^2 - 2u + v^2 - \frac{2v}{x} = -1$$

add: $r^2/(r+1)^2$ to both sides   add $1/x^2$ to both sides

$$u^2 - \frac{2ru}{r+1} + \frac{r^2}{(r+1)^2} + v^2$$

$$= \frac{1-r}{1+r} + \frac{r^2}{(1+r)^2}$$

$$u^2 - 2u + 1 + v^2 - \frac{2r}{x} + \frac{1}{x^2}$$

$$= \frac{1}{x^2}$$

$$\left(u - \frac{r}{r+1}\right)^2 + v^2 = \frac{1}{(1+r)^2}$$

$$(u-1)^2 + \left(v - \frac{1}{x}\right)^2 = \frac{1}{x^2}$$

Therefore two sets of circles are produced with the following radii and centres.

Real part of Z   **Imaginary part of Z**

$$\text{Radius} = \frac{1}{1+r}$$ $$\text{Radius} = \frac{1}{x}$$ (3.114a/b)

$$\text{Centre} = \frac{r}{1+r} : 0$$ $$\text{Centre} = 1 : \frac{1}{x}$$ (3.115a/b)

The circles of constant real impedance are shown in Figure 3.19a and the circles of constant imaginary impedance are shown in Figure 3.19b. A Smith Chart is shown in Figure 3.20. The S parameters can be directly overlaid onto this and the impedance deduced and vice versa. This will be illustrated under amplifier matching using S parameters.

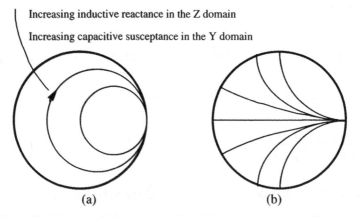

Increasing inductive reactance in the Z domain

Increasing capacitive susceptance in the Y domain

(a)   (b)

Figure 3.19 Circles of (a) constant real and (b) constant imaginary impedance

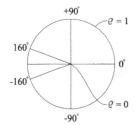

IMPEDANCE OR ADMITTANCE COORDINATES

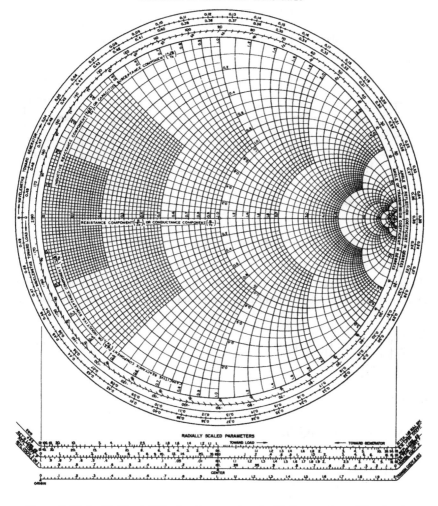

Figure 3.20 Smith Chart. This chart is reproduced with the courtesy of the Analog
Instrument Co., Box 950, New Providence, NJ 07974, USA

3.7.2 Input and Output Impedance

Most RF measurements above 5 MHz are now performed using S parameter network analysers and therefore amplifier design using S parameters will be discussed. In the amplifier it is important to obtain the input and output impedance/reflection coefficient. An S parameter model of an amplifier is shown in Figure 3.21. Here impedances will be expressed in terms of reflection coefficients normalised to an impedance Z_0. These are Γ_{in}, Γ_{out}, Γ_S and Γ_L.

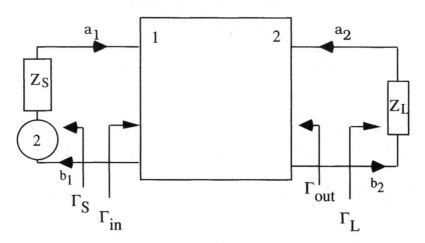

Figure 3.21 Two port S parameter model

Taking the S parameter two port matrix the input and output reflection coefficients can be derived which offers significant insight into stability and error correction.

$$\begin{pmatrix} b_1 \\ b_2 \end{pmatrix} = \begin{pmatrix} S_{11} & S_{12} \\ S_{21} & S_{22} \end{pmatrix} \begin{pmatrix} a_1 \\ a_2 \end{pmatrix} \tag{3.116}$$

Therefore:

$$b_1 = S_{11}a_1 + S_{12}a_2 \text{ and } b_2 = S_{21}a_1 + S_{22}a_2 \tag{3.117}$$

$$\Gamma_{in} = \frac{b_1}{a_1} = S_{11} + S_{12}\frac{a_2}{a_1} \tag{3.118}$$

as:

$$a_1 = \frac{b_2 - S_{22}a_2}{S_{21}} \tag{3.119}$$

$$\Gamma_{in} = S_{11} + S_{12}S_{21}\left(\frac{a_2}{b_2 - S_{22}a_2}\right) \tag{3.120}$$

dividing by b_2:

$$\Gamma_{in} = S_{11} + S_{12}S_{21}\left(\frac{\Gamma_L}{1 - S_{22}\Gamma_L}\right) \tag{3.121}$$

Similarly the reflection coefficient from the other port is:

$$\Gamma_{out} = \frac{b_2}{a_2} = S_{22} + S_{12}S_{21}\left(\frac{\Gamma_S}{1 - S_{11}\Gamma_S}\right) \tag{3.122}$$

It can be seen, in the same way as for the y parameters, that the input reflection coefficient is dependent on the load and that the output reflection coefficient is dependent on the source. These equations are extremely useful for the calculation of stability where the dependence of input and output reflection coefficient with load and source impedance respectively is analysed. Similarly these equations are also used to calculate error correction by modelling the interconnecting cable as a two port network. Note that Γ_{in} and Γ_{out} become S_{11} and S_{22} respectively when the load and source impedances are Z_0. This also occurs when $S_{12} = 0$ in the case of no feedback.

3.7.3   Stability

For stability it is required that the magnitude of the input and output reflection coefficient does not exceed one. In other words, the power reflected is always lower than the incident power. For unconditional stability the magnitude of the input and output reflection coefficients are less than one for all source and loads whose magnitude of reflection coefficients are also less than one. For unconditional stability it is therefore required that:

$$|\Gamma_{in}| < 1 \tag{3.123}$$

$$|\Gamma_{out}| < 1 \tag{3.124}$$

for all $|\Gamma_S| < 1$ and for all $|G_L| < 1$.

Examining, for example, the equation for input reflection coefficient it is possible to make some general comments about what could cause instability such that $|\Gamma_{in}| < 1$. If the product of $S_{12}S_{21}$ is large then there is a strong possibility that certain load impedances Γ_L could cause instability. As S_{21} is usually the required gain it is therefore important to have good reverse isolation such that S_{12} is low. If the input match is poor such that S_{11} is large then the effect of the second term is therefore even more important. This also illustrates how a circuit can be forced to be unconditionally stable by restricting the maximum value of Γ_L. This is most easily achieved by placing a resistor straight across the output. In fact both shunt and series resistors can be used. It is also very easy to calculate the value of the resistor directly from the S parameters and the equation for Γ_{in} and Γ_{out}.

Take, for example, a simple stability calculation for a device with the following S parameters. For convenience parameters with no phase angle will be chosen. Assume that $S_{11} = 0.7$, $S_{21} = 5$, $S_{12} = 0.1$ and $S_{22} = 0.2$. Calculating Γ_{in}:

$$\Gamma_{in} = S_{11} + S_{12}S_{21}\left(\frac{\Gamma_L}{1 - S_{22}\Gamma_L}\right) = 0.7 + 0.5\left(\frac{\Gamma_L}{1 - 0.2\Gamma_L}\right) \tag{3.125}$$

It can be seen that if:

$$\left(\frac{\Gamma_L}{1 - 0.2\Gamma_L}\right) > 0.6 \tag{3.126}$$

then the equation for Γ_{in} predicts instability as Γ_{in} exceeds one. This occurs when $\Gamma_L(1 + 0.12) = 0.6$, therefore $\Gamma_L = 0.536$.

Therefore the resistor value is:

$$Z_L = Z_0 \frac{1 + \rho}{1 - \rho} = 165\Omega \tag{3.127}$$

This means that if a shunt resistor of 165Ω or less was placed across the output then the input reflection coefficient could never exceed one even if the circuit was connected to an open circuit load.

Note also that the same checks must be done for the output reflection coefficient Γ_{out}. It is therefore left to the reader to check whether there are any requirements for the source impedance to ensure stability. It should be noted that it is usually preferable to place resistive components across the output rather than the input as loss at the input degrades the noise figure.

It is also essential to calculate stability at all usable frequencies including those outside the band of operation as the source and load impedances are often unpredictable. This is usually done using CAD tools. A number of factors have therefore been developed. For unconditional stability the Rollett stability factor is stated as:

$$K = \frac{1 - |S_{11}|^2 - |S_{22}|^2 + |S_{11}S_{22} - S_{12}S_{21}|^2}{2S_{12}S_{21}} > 1 \qquad (3.128)$$

The Stern factor [14] is $1/K$ and also $|S_{11}S_{22} - S_{12}S_{21}| < 1$ $\qquad$ (3.129)

Stability can also be demonstrated using the Smith Chart by plotting the loci of points of the load and source reflection coefficients (Γ_L and Γ_s) which produce $|\Gamma_{in}| = 1$ and $|\Gamma_{out}| = 1$. These loci [1] consist of circles with a centre C_L and radius r_L as shown in the following equations and Figure 3.22.

Γ_L values for $|\Gamma_{in}| = 1$ Γ_S values for $|\Gamma_{out}| = 1$

(output stability circle)   (input stability circle)

$$r_L = \left| \frac{S_{12}S_{21}}{|S_{22}|^2 - |\Delta|^2} \right| \quad \text{radius} \qquad r_S = \left| \frac{S_{12}S_{21}}{|S_{11}|^2 - |\Delta|^2} \right| \quad \text{radius} \qquad (3.130a/b)$$

$$C_L = \frac{\left(S_{22} - \Delta S_{11}^*\right)^*}{|S_{22}|^2 - |\Delta|^2} \quad \text{centre} \qquad C_S = \frac{\left(S_{11} - \Delta S_{22}^*\right)^*}{|S_{11}|^2 - |\Delta|^2} \quad \text{centre} \qquad (3.131a/b)$$

where: $\Delta = S_{11}\,S_{22} - S_{12}\,S_{21}$ $\qquad\qquad\qquad\qquad$ (3.122)

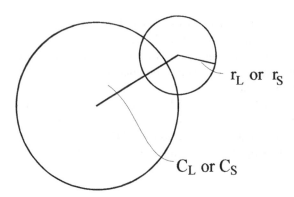

Figure 3.21 Stability Circles

It is necessary to check which part of the circle is the stable region. This is achieved by letting $Z_L = Z_0$ causing Γ_L to be zero and Γ_{in} to be S_{11}. Similarly by letting $Z_S = Z_0$, this causes Γ_S to be zero and Γ_{out} to be S_{22}. These points ($Z_L = Z_0$) are the centre of the Smith Chart. If the device is stable with source and load impedances equal to Z_0 (i.e. $|S_{11}| < 1$) then the centre of the Smith Chart is a stable point. This is most easily illustrated by looking at Figure 3.23.

The devices are unconditionally stable when the circles do not overlap (Figure 3.23b). When Γ_L is zero but $|S_{11}| > 1$ then the device is unstable even when terminated in Z_0, the device cannot be unconditionally stable. However, it can be conditionally stable where the two circles overlap within the Smith Chart as shown in Fig 3.23c. The gain of the amplifier can now be considered in terms of the source and load impedances and the device S parameters.

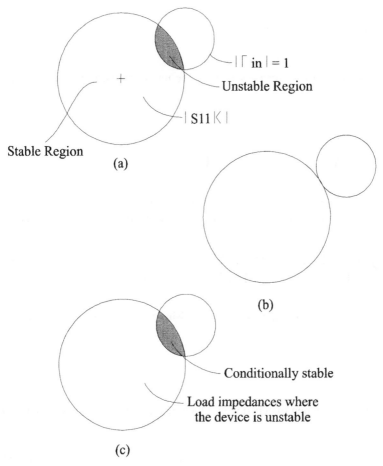

$|\Gamma \text{ in}| = 1$

Unstable Region

$|S11 < |$

Stable Region

(a)

(b)

Conditionally stable

Load impedances where
the device is unstable

(c)

Figure 3.23 Stability Circles demonstrating the stable regions

3.7.4   Gain

As for y parameters, let us investigate the transducer gain, G_T, in terms of S parameters:

$$G_T = \frac{P_L}{P_{AVS}} = \frac{1-\left|\Gamma_S\right|^2}{\left|1-\Gamma_{in}\Gamma_S\right|^2} \left|S_{21}\right|^2 \frac{1-\left|\Gamma_L\right|^2}{\left|1-S_{22}\Gamma_L\right|^2} \tag{3.123}$$

Note that:

$$\Gamma_{in} = S_{11} - S_{12}\frac{a_2}{a_1} \qquad\qquad (3.124)$$

If S_{12} is assumed to be zero, by making the unilateral assumption, then $\Gamma_{in} = S_{11}$ and therefore:

$$G_{TU} = \frac{1-|\Gamma_S|^2}{|1-S_{11}\Gamma_S|^2}\ |S_{21}|^2\ \frac{1-|\Gamma_L|^2}{|1-S_{22}\Gamma_L|^2} \qquad\qquad (3.125)$$

The maximum unilateral gain, *MUG* or *GUM*, occurs when: $\Gamma_S = S_{11}^*$ and $\Gamma_L = S_{22}^*$ where the asterisk is the complex conjugate. The equation for *MUG* is therefore:

$$MUG = \frac{1}{1-|S_{11}|^2}\ |S_{21}|^2\ \frac{1}{1-|S_{22}|^2} \qquad\qquad (3.126)$$

This can be most easily understood by realising that when the input is not matched P_{in} is:

$$P_{in} = 1 - |S_{11}|^2 \qquad\qquad (3.127)$$

The input power when matched would then be increased by the reciprocal of P_{in}:

$$\frac{1}{1-|S_{11}|^2} \qquad\qquad (3.128)$$

The same argument applies to the output.

3.7.4.1   Other Gains

The power gain is:

$$G_P = \frac{P_L}{P_{in}} \qquad\qquad (3.129)$$

$$P_{in} = |a_1|^2 - |b_1|^2 = |a_1|^2 \left(1 - |\Gamma_{in}|^2\right) \tag{3.130}$$

$$G_P = \frac{1}{1 - |\Gamma_{in}|^2} \, |S_{21}|^2 \, \frac{1 - |\Gamma_L|^2}{|1 - S_{22}\Gamma_L|^2} \tag{3.131}$$

The available power gain is:

$$G_A = \frac{P_{AVN}}{P_{AVS}} \tag{3.132}$$

$$G_A = \frac{1 - |\Gamma_S|^2}{|1 - S_{11}\Gamma_S|^2} \, |S_{21}|^2 \, \frac{1}{1 - |\Gamma_{out}|^2} \tag{3.133}$$

3.7.5   Simultaneous Conjugate Match

When S_{12} is not equal to zero or is rather large the unilateral assumption cannot be made and the input and output reflection coefficients are given by:

$$\Gamma_{in} = S_{11} + \frac{S_{12} S_{21} \Gamma_L}{1 - S_{22}\Gamma_L} \tag{3.134}$$

$$\Gamma_{out} = S_{22} + \frac{S_{12} S_{21} \Gamma_S}{1 - S_{11}\Gamma_S} \tag{3.135}$$

The conditions required to obtain maximum transducer gain then require that both the input and output should be matched simultaneously. Therefore:

$$\Gamma_S = \Gamma^*_{in} \tag{3.136}$$

and

$$\Gamma_L = \Gamma^*_{out} \tag{3.137}$$

Therefore:

$$\Gamma^*_S = S_{11} + \frac{S_{12} S_{21} \Gamma_L}{1 - S_{22}\Gamma_L} \tag{3.138}$$

and:

$$\Gamma^*_L = S_{22} + \frac{S_{12} S_{21} \Gamma_S}{1 - S_{11}\Gamma_S} \tag{3.139}$$

It should be noted that the output matching network and load impedance affects the input impedance and the input matching network and source impedance affects the output impedance. It is, however, possible with considerable manipulation to derive equations for this simultaneous match condition as long as the device is stable. These are shown below [1]:

$$\Gamma_{Sopt} = \frac{B_1 \pm \left(B_1^2 - 4|C_1|^2\right)^{1/2}}{2C_1} \tag{3.140}$$

$$\Gamma_{Lopt} = \frac{B_2 \pm \left(B_2^2 - 4|C_2|^2\right)^{1/2}}{2C_2} \tag{3.141}$$

$$B_1 = 1 + |S_{11}|^2 - |S_{22}|^2 - |\Delta|^2 \tag{3.142}$$

$$B_2 = 1 - |S_{11}|^2 + |S_{22}|^2 - |\Delta|^2 \tag{3.143}$$

$$C_1 = S_{11} - S_{22}^*\Delta \tag{3.144}$$

$$C_2 = S_{22} - S_{11}^*\Delta \tag{3.145}$$

$$\Delta = S_{11}S_{22} - S_{12}S_{21} \tag{3.146}$$

Now that the input and output impedances and the gain equations are known the amplifier matching networks can be designed using the Smith Chart. For these calculations S_{12} will be assumed to be zero and therefore the input reflection coefficient is S_{11} and the output reflection coefficient is S_{22}.

3.7.6   Narrow Band Matching Using the Smith Chart for Unilateral Amplifier Design

Matching the input and output of the device uses the same procedure. The following rules should be followed.

1.   Normalise the characteristic impedance of the Smith Chart to Z_0. Note that the centre point of the Smith Chart is 1. When working in the impedance domain, divide by Z_0 to place the point on the chart and multiply the data when removing from the chart. For example, place the impedance (80 + 20j) on a 50Ω Smith Chart. As this is 50(1.6 + 0.4j) place the point 1.6 + 0.4j on the chart. In the admittance domain multiply by Z_0 to place on the chart and divide by Z_0 to remove from the chart.

2.   Plot the device impedance Z on the Smith Chart either directly if known as impedance or as S_{11} or S_{22}. If the impedance is in S parameters note that S_{11} is a reflection coefficient which is the polar form of the Smith Chart. The outer circle of the Smith Chart has a reflection coefficient of 1 ($\rho = 1$) and the centre of the Smith Chart has $\rho = 0$. The magnitude of ρ varies linearly between 0 and 1 along a radius , R, of the chart.

3.   Using a ruler measure the radius (R) between the centre and the outer circumference of the Smith Chart and then draw a circle, from the centre, of radius of ρR. For example, if the radius of the circle is 10cm and the magnitude of $\rho = 0.3$, draw a circle with a radius of 3cm. Using the angles on the Smith Chart draw a line originating from the centre of the chart through the circle intersecting the angle. The angle is zero at the RHS of the Smith Chart and increases positively in an anti-clockwise direction.

It is now necessary, by the use of series and parallel components, to transform the input or output impedance of the device to Z_0 by moving towards the generator. Note that the observer is always looking towards the device and while moving through the matching network moves to the middle of the Smith Chart to produce a match. A block diagram of an amplifier consisting of the device and the matching components is shown in Figure 3.24

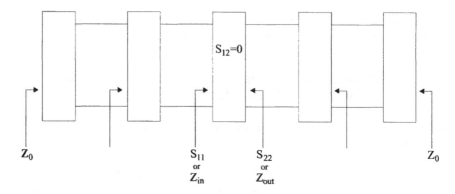

Figure 3.24 Block diagram of amplifier matching

3.7.7   LC Matching Networks

The Smith Chart will be used here to design a two reactive element, L type, matching network which consist of a series and shunt reactance as shown in Figure 3.25.

The operation of matching, by the use of series and shunt components, can be illustrated generically. When an impedance a + jb is to be matched to a real impedance D using a series impedance and shunt reactance, the series component is added to produce a + jb + jc such that the admittance of this (inverse of the impedance) provides the correct real part of 1/D. This is because the final shunt component can only change the imaginary part of the admittance as it is a pure reactance. The imaginary part is then cancelled by the addition of a shunt component of opposite reactance.

The L matching network can have either the shunt component first or the series component first. The Smith Chart can be represented as an admittance chart by flipping around a vertical line through the centre to produce a mirror image. It is possible to obtain charts with curves for constant real and imaginary admittance overlaid on the same chart but these can become very confusing. Therefore a better way to convert from impedances to admittances and vice versa is obtained by drawing a line from the known impedance/admittance through the centre of the chart and extending equally to the other side. This can also be helped by adding Z = 1 and Y = 1 circles as illustrated in the following examples. When series impedances are added, the Smith Chart is operated as an impedance chart as impedances just add, and when a shunt impedance/admittance is being added the Smith Chart is operated in the admittance domain as the ådmittances just add. It is, however, important to remember which domain you are working in at any one time.

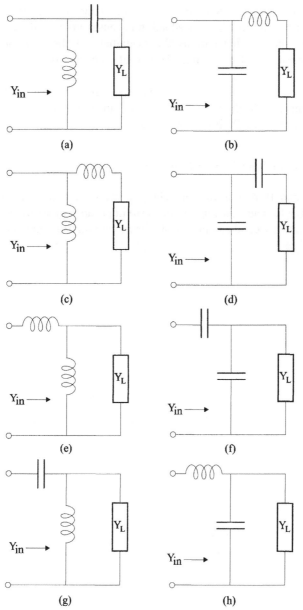

Figure 3.25 *LC* matching networks

3.7.8   Transmission Line Matching Networks

When a series transmission line, with the same normalised characteristic impedance of the Smith Chart, is added, this causes the impedance to move in a circle around the Smith Chart as the magnitude of the reflection coefficient remains constant for a lossless line. This would typically be moved until it meets the unity admittance line.

For shunt stubs the value of susceptance required is obtained and then an open or short circuit length of line is used which is measured as a length of line with $|\rho| = 1$. This is most easily illustrated by four examples.

3.7.9   Smith Chart Design Examples

Example 1a and 1b demonstrate matching to the same impedance using inductor capacitor, *LC*, (1a) and transmission line (1b) impedance transformers. Example 2a and 2b illustrate a second example again for *LC* and transmission line networks.

Example 1a: Using the Smith Chart shown in Figure 3.26 match an impedance with input reflection coefficient $\rho = 0.4$ angle $+136°$ to 50Ω using LC matching networks.

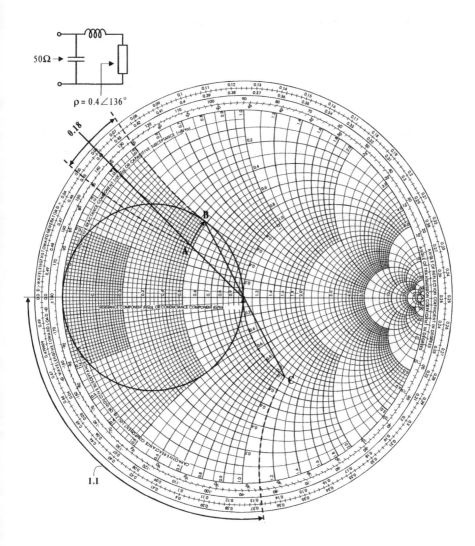

Figure 3.26 Example 1a: Match $\rho = 0.4$ angle $+136°$ degrees to 50Ω

Example 1.a: Using the Smith Chart shown in Figure 3.26 match an impedance with input reflection coefficient $\rho = 0.4$ angle $+136°$ to 50Ω using LC matching networks.

1.   Measure the radius (R) of the Smith Chart.

2.   Plot the magnitude $R.\rho$ on the Smith Chart at A. Note that many charts have a scale for ρ below the chart. This varies linearly from 0 to 1 over one radius of the chart and can therefore be used with the aid of a compass.

3.   Draw a circle on the Smith Chart which is the mirror image about a central vertical line of the Z = 1 circle. (This enables the impedance to be converted to admittance at the correct point.)

4.   Add inductive reactance to move A to this circle (point B). Measure the change in reactance.

5.   This is jX. As $Z_0jX = j\omega L$ therefore $L = Z_0 X/\omega$.

6.   Convert to admittance (C) and add shunt capacitive susceptance to bring to the middle of the chart: $X.Y_0 = j\omega C$ and $jX/Z_0 = j\omega C$.

Note that if A was outside the circle (the mirror image of the Z = 1 circle) then the first component would need to be a series capacitor.

Example 1b: Using the Smith Chart shown in Figure 3.27 match an impedance with input reflection coefficient $\rho = 0.4$ angle $+136°$ to 50Ω using transmission line matching networks.

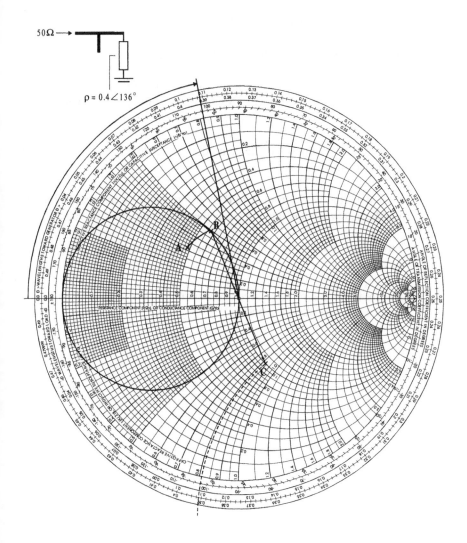

Figure 3.27 Example 1b: Match $\rho = 0.4$ angle $+136°$ to 50Ω

Example 1b: Using the Smith Chart shown in Figure 3.27 match an impedance with input reflection coefficient $\rho = 0.4$ angle $+136°$ to 50Ω using transmission line matching networks.

1.   Measure the radius (R) of the Smith Chart.

2.   Plot the magnitude $R.\rho$ on the Smith Chart at A.

3.   Draw a circle on the Smith Chart which is the mirror image about a central vertical line of the Z=1 circle. (This enables the impedance to be converted to admittance at the correct point.) This produces a Y=1 circle.

4.   Using a compass draw a circle centred at the middle of the Smith Chart and intersecting point A.

5.   Move along the circle from A to B by using a series transmission line. B should occur where the line intersects the admittance circle of value 1.

6.   Convert the chart to an admittance chart and measure the value of the susceptance.

7.   Measure the length of open circuit or short circuit line required to produce an equal and opposite susceptance.

Example 2a: Using the Smith Chart shown in Figure 3.28 design an LC matching circuit to match (80Ω + 20j) to 50Ω.

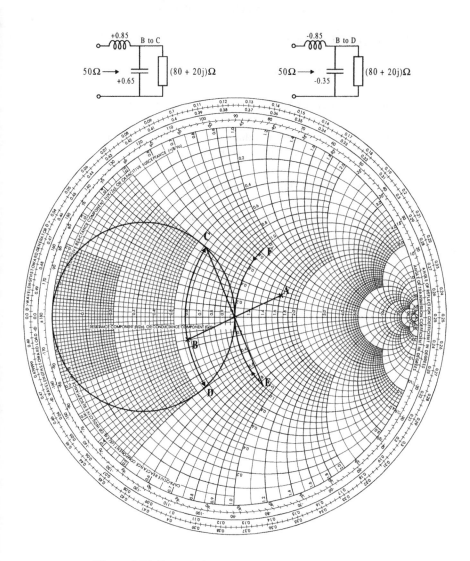

Figure 3.28 Example 2a: Match Z = (80Ω +20j) to 50 ohms.

Example 2a: Using the Smith Chart shown in Figure 3.28:

1.   Normalise to 50Ω, Z = (80 + 20j)/50 = 1.6 +0.4j.

2.   Plot on Smith Chart at point A.

3.   Convert to admittance at B.

4.   Draw mirror circle of Y = 1 circle to produce Z = 1 circle.

5.   Add shunt capacitance or inductance to move to C or D respectively
        where $-jX/Z_0 = 1/j\omega L = -j/\omega L = -jX.Y_0$ or $jX/Z_0 = j\omega C$.

6.   Convert back to impedance domain E or F.

7.   Add series reactance or susceptance to obtain match.

8. $Z_0jX = j\omega$ or $-Z_0jX = 1/j\omega c = -j/\omega C$.

Example 2.b: Using the Smith Chart shown in Figure 3.29, design a transmission line matching circuit to match (80Ω + 20j) to 50Ω.

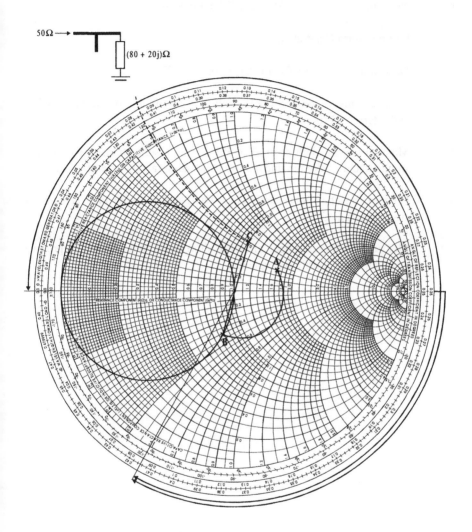

Figure 3.29 Example 2b: Match Z = (80Ω +20j) to 50Ω

Example 2b: Using the Smith Chart shown in Figure 3.29.

1.   Normalise to 50Ω, $Z = (80 + 20j)/50 = 1.6 + 0.4j$.

2.   Plot on Smith Chart at A.

3.   Draw mirror circle of $Z = 1$ circle to produce $Y = 1$ circle.

4.   Using a compass draw a circle centred at the middle of the Smith Chart and intersecting point A.

5.   Move along the circle from A to B by using a series transmission line. B should occur where the line intersects the admittance circle of value 1.

6.   Convert the chart to an admittance chart and measure the value of the susceptance at C.

7.   Measure the length of open circuit or short circuit line required to produce an equal and opposite susceptance.

3.7.10   Amplifier Problems

1.   Design an amplifier using LC circuits and a Smith Chart operating at 1 GHz based on the following S parameters.

$$S_{11} = 0.6\angle180°$$
$$S_{22} = 0.5\angle180°$$
$$S_{21} = 3$$
$$S_{12} = 0.00005$$

Calculate the Maximum Unilateral Gain (MUG).

2.   Design an amplifier using LC circuits and a Smith Chart operating at 1 GHz based on the following S parameters.

$$S_{11} = 0.9\angle0°$$
$$S_{22} = 0.5\angle0°$$
$$S_{21} = 3$$
$$S_{12} = 0.00005$$

2.   Calculate whether a device driven from a 50Ω source with the following S parameters is stable into a 50Ω load and a 75Ω load.

$$S_{11} = 0.9\angle0°$$
$$S_{22} = 0.5\angle0°$$
$$S_{21} = 5\angle0°$$
$$S_{12} = 0.2\angle0°$$

3.8   Broadband Feedback Amplifiers

There is often a requirement to design broadband amplifiers with good input and output match. This section describes the design of broadband feedback amplifiers of the form shown in Figure 3.30 which includes either a feedback resistor R_F or both a feedback resistor R_F and an emitter feedback resistor R_E. It will be shown that suitable combinations of resistors can be chosen to produce broadband design with low return loss such that S_{11} and S_{22} tend to zero.

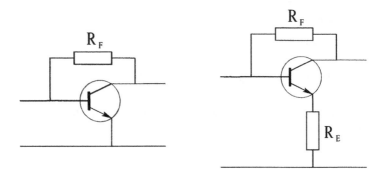

Figure 3.30 Topology for broadband amplifiers

To enable easy analysis of these circuits it is worth investigating the low frequency T and π models for the bipolar transistor. To ease analysis it is also worth assuming that the current gain β is infinite; therefore the π model is now modified to have an infinite input impedance. The external emitter resistor R_E can also be incorporated into this π model by using a modified g_m defined as $g_m{'}$ which includes both the emitter AC resistance and the external resistor R_E such that:

$$g_m{'} = \frac{1}{r_e + R_E} = \frac{1}{\dfrac{1}{g_m} + R_E}$$

(3.147)

Therefore:

$$g_m{'} = \frac{g_m}{1 + g_m R_E}$$

(3.148)

The model, including the feedback resistor, R_F, and modified g_m' is shown in Figure 3.31.

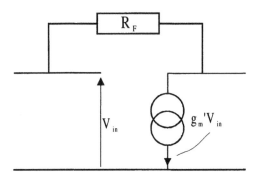

Figure 3.31 Simple model for feedback amplifier

It is now worth calculating the input and output impedance of the amplifier and S_{21} and S_{12}. This also relates to the section on the 'generic Miller effect' described in chapter 2 where it was shown that the input impedance of an amplifier with a feedback resistor is reduced by the ratio $R_F/(1 + G)$. Taking the model shown in Figure 3.32 the output impedance can be derived by calculating the ratio of V_x/I_x.

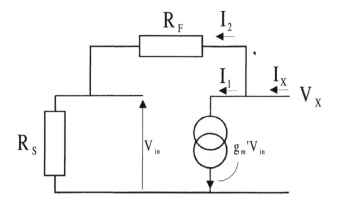

Figure 3.32 Model for calculating the output impedance

As:

$$I_X = I_1 + I_2 \tag{3.149}$$

and:

$$I_1 = g_m{}'V_{in}$$ (3.150)

and:

$$I_2 = \frac{V_X - V_{in}}{R_F}$$ (3.151)

As

$$V_{in} = V_X \frac{R_S}{R_S + R_F}$$ (3.152)

I_1 is therefore:

$$I_1 = g_m{}'V_X \frac{R_S}{R_S + R_F}$$ (3.153)

and I_2 is therefore:

$$I_2 = \frac{V_X}{R_F}\left(1 - \frac{R_S}{R_S + R_F}\right)$$ (3.154)

The total current I_X is therefore:

$$I_X = I_1 + I_2 = V_X\left(g_m{}'\frac{R_S}{R_S + R_F} + \frac{1}{R_F} - \frac{R_S}{R_F(R_S + R_F)}\right)$$ (3.155)

The output impedance is therefore:

$$\frac{V_X}{I_X} = \frac{1}{\left(g_m{}'\dfrac{R_S}{R_S + R_F} + \dfrac{1}{R_F} - \dfrac{R_S}{R_F(R_S + R_F)}\right)}$$ (3.156)

Making $R_F(R_S + R_F)$ the common factor in the denominator:

$$\frac{V_X}{I_X} = \cfrac{1}{\cfrac{R_S + R_F - R_S + g_m'R_SR_F}{R_F(R_S + R_F)}}$$

(3.157)

The output impedance is therefore:

$$\frac{R_S + R_F}{1 + g_m'R_S}$$

(3.158)

The input impedance can be calculated in a similar way by calculating V_{in}/I_{in} using the model shown in Figure 3.33.

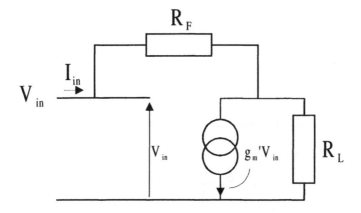

Figure 3.33 Model for calculating input impedance

Calculating V_{out} in terms of V_{in}:

As:

$$I_{in} = \frac{V_{in} - V_{out}}{R_F}$$

(3.159)

then:

$$V_{out} = -g_m{'}V_{in}R_L + \left(\frac{V_{in} - V_{out}}{R_F}\right)R_L \tag{3.160}$$

$$V_{out}\left(1 + \frac{R_L}{R_F}\right) = -g_m{'}V_{in}R_L + V_{in}\frac{R_L}{R_F} \tag{3.161}$$

Therefore:

$$V_{out}\left(R_F + R_L\right) = V_{in}R_L\left(1 - g_m{'}R_F\right) \tag{3.162}$$

$$V_{out} = V_{in}R_L\frac{\left(1 - g_m{'}R_F\right)}{\left(R_F + R_L\right)} \tag{3.163}$$

As:

$$I_{in} = \frac{V_{in} - V_{out}}{R_F} \tag{3.164}$$

then the input impedance is therefore:

$$\frac{V_{in}}{I_{in}} = \frac{V_{in}R_F}{V_{in}\left(1 - \frac{R_L\left(1 - g_m{'}R_F\right)}{R_F + R_L}\right)} \tag{3.165}$$

$$\frac{V_{in}}{I_{in}} = \frac{R_F\left(R_F + R_L\right)}{R_F + R_L - R_L + g_m{'}R_F R_L} \tag{3.166}$$

$$\frac{V_{in}}{I_{in}} = \frac{R_F + R_L}{1 + g_m{'}R_L} \tag{3.167}$$

Note therefore that if the source and load impedances are Z_0 then both the input and output impedance is therefore:

$$\frac{R_F + Z_0}{1 + g_m'Z_0} \tag{3.168}$$

S_{21} can be calculated using the 2 Volt rule described in chapter 2 by calculating the output voltage of the model shown in Figure 3.34 and by assuming that $R_S = R_L = Z_0$.

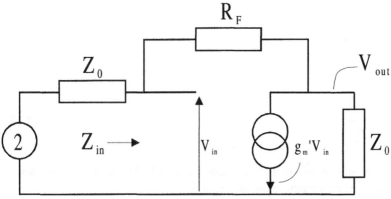

Figure 3.34 Model for calculating S_{21} for amplifier

To obtain S_{21}, calculate V_{in} in terms of the input impedance Z_{in}. by substituting equation (3.168) for Z_{in} as follows:

$$V_{in} = \frac{2Z_{in}}{Z_{in} + Z_0} = \frac{2\left(\dfrac{R_F + Z_0}{1 + g_m'Z_0}\right)}{\dfrac{R_F + Z_0}{1 + g_m'Z_0} + Z_0} \tag{3.169}$$

therefore:

$$V_{in} = \frac{2(R_F + Z_0)}{R_F + Z_0 + Z_0(1 + g_m'Z_0)} \tag{3.170}$$

substituting this in equation (3.163) and making $R_L = Z_0$ then as $S_{21} = V_{out}$:

$$S_{21} = \frac{2(R_F + Z_0)}{R_F + 2Z_0 + g_m'Z_0^2} \times Z_0 \frac{(1 - g_m'R_F)}{(R_F + Z_0)} \qquad (3.171)$$

Taking equation (3.168) and assuming now that both the input and output impedance are equal to Z_0 such that S_{11} and S_{22} equal zero, then:

$$Z_0 = \frac{R_F + Z_0}{1 + g_m'Z_0} \qquad (3.172)$$

Therefore:

$$g_{m'} = \frac{R_F}{Z_0^2} \qquad (3.173)$$

If this is substituted in the equation for S_{21} (3.171) then the first part of the expression is now equal to one and S_{21} is now:

$$S_{21} = \frac{Z_0 - \dfrac{R_F^2}{Z_0}}{R_F + Z_0} \qquad (3.174)$$

Multiplying top and bottom by Z_0:

$$S_{21} = \frac{Z_0^2 - R_F^2}{R_F Z_0 + Z_0^2} \qquad (3.175)$$

$$S_{21} = \frac{(Z_0 - R_F)(Z_0 + R_F)}{Z_0(Z_0 + R_F)} \qquad (3.176)$$

which simplifies to:

$$S_{21} = \frac{Z_0 - R_F}{Z_0} \qquad (3.177)$$

R_F is therefore:

$$R_F = Z_0(1 - S_{21})$$ (3.178)

Note that a simpler way of calculating S_{21} would have been to assume that the input is matched from the start. Therefore using the 2 volt rule, V_{in} would be one which would have reduced the number of equations as S_{21} is just V_{out}. S_{12} can therefore be calculated more easily. Assume the output impedance is Z_0 the voltage applied at the output would be 1 volt and therefore the voltage across the source resistor Z_0 would be S_{12}. By potential division S_{12} is therefore:

$$S_{12} = \frac{Z_0}{Z_0 + R_F}$$ (3.179)

If no external emitter resistor, R_E, is to be used then g_m must be set by the current:

$$g_m' = g_m = \frac{R_F}{Z_0^2} = \frac{(1 - S_{21})}{Z_0}$$ (3.180)

Remember that S_{21} is negative as the device is inverting.

3.8.1   Broadband Design Examples

Take an example of a 50Ω amplifier with a gain, $(S_{21})^2$ of 10dB.

$$R_F = 50(1 + \sqrt{10}) = 208\Omega$$ (3.181)

The transconductance g_m is therefore:

$$g_m = \frac{(1 + \sqrt{10})}{50} = 0.083$$ (3.182)

Remembering that:

$$r_e = \frac{1}{g_m} = \frac{25mV}{I_E}$$ (3.183)

the current would therefore be 2.1 mA as $I_E = 25*10^{-3}$ gm.

Now examine the case where an emitter resistor R_E is also to be included. Taking the equation for g_m' (3.148):

$$g_m' = \frac{g_m}{1 + g_m R_E} = \frac{R_F}{Z_0^2} \tag{3.148}$$

$$R_E = \frac{Z_0^2}{R_F} - \frac{1}{g_m} = \frac{Z_0^2}{R_F} - r_e \tag{3.184}$$

Take another example. Design a broadband amplifier with 12dB gain operating at 10mA, then with terminating impedances of 50Ω.

$$R_F = 50\left(1 + \sqrt{15.8}\right) = 250\Omega \tag{3.185}$$

$$R_E = \frac{50^2}{250} - 2.5 = 7.5\Omega \tag{3.186}$$

S_{12} would be 50/300 and is therefore around –16dB. An amplifier based on these equations is shown in Figure 3.35. Here a BFG520 transistor is biased to 10mA with two 15Ω resistors in parallel in the emitter and a 270Ω feedback resistor. Biasing resistors are also included.

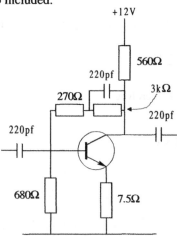

Figure 3.35 50Ω broadband feedback amplifier

S_{21} and S_{11} are shown in Figure 3.36 and S_{12} and S_{22} in Figure 3.37. This circuit contains no high frequency gain compensation. Compensation can be used to improve the high frequency response as shown in Gonzalez [1] and Hayward [2] where a small inductance can be added in series with the feedback resistor R_F, the emitter resistor R_E or the output. The effect of the emitter contact resistor of around 1Ω was ignored in this design which is why the gain is slightly low and this could easily be included. S_{11} is better that -20dB up to 1.5 GHz and S_{22} starts at —15dB degrading as the frequency is increased. S_{12} is as predicted at around −16dB.

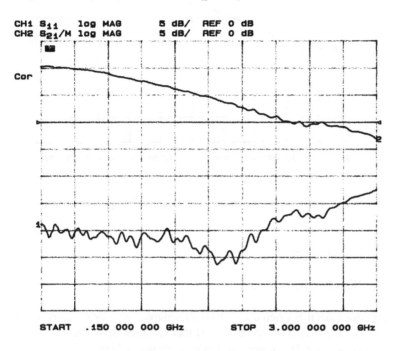

Figure 3.36 S_{21} and S_{11} for Broadband Amplifier

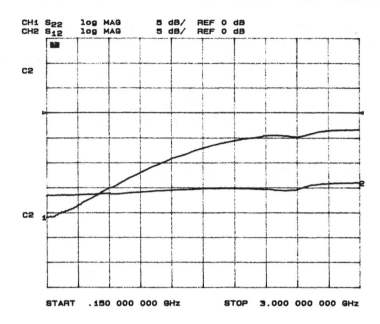

Figure 3.37 S_{21} and S_{11} for Broadband Amplifier

3.9   DC Biasing

3.9.1   Bipolar Transistors

At low frequencies it is often useful to use decoupled emitter resistors and a fairly
high base voltage bias to reduce the effect that variation in V_{be} has on the collector
current. A typical low frequency bias circuit is shown in Figure 3.38.

At higher frequencies the emitter components often cause instability and may
increase the noise. It is therefore important to develop biasing circuits to
compensate for the device to device variation in β and for the temperature
variation of V_{be}.

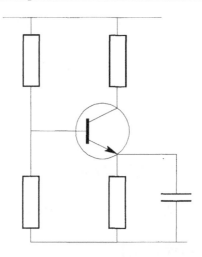

Figure 3.28 LF Bias Circuit

This can be achieved by placing a resistor in the collector circuit which is used to reduce the base bias as the current increases. This is illustrated in Figure 3.39.

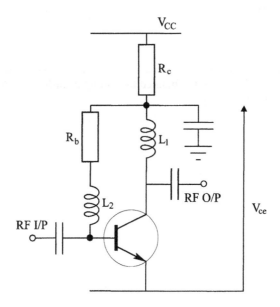

Figure 3.39 RF Bias Circuit

Take an example: Design a bias circuit offering I_C = 5mA, V_{CE} = 5V and V_{CC} = 15V. ThereforeR_C = 2kΩ. Assume β = 50. As $I_C = \beta i_b$

$$V_{CE} - 0.7 = i_b R_b \tag{3.187}$$

$$R_b = \frac{V_{ce} - 0.7}{i_b} = \frac{4.3}{10^{-4}} = 43\text{k}\Omega \tag{3.188}$$

$$V_{cc} - V_{ce} = \left(I_c + i_b\right)R_c = \left(\beta + 1\right)i_b R_c \tag{3.189}$$

$$V_{cc} = \left(\beta + 1\right)i_b R_c + i_b R_b + 0.7 = i_b\left[\left(\beta + 1\right)R_c + R_b\right] + 0.7 \tag{3.190}$$

$$V_{cc} - 0.7 = \frac{I_c\left[\left(\beta + 1\right)R_c + R_b\right]}{\beta} \tag{3.191}$$

Therefore:

$$I_c = \frac{\beta\left(V_{cc} - 0.7\right)}{\left[\left(\beta + 1\right)R_c + R_b\right]} \tag{3.192}$$

If β increases from 50 to 100 then I_c changes from 5mA to 5.83mA.

Lower resistor values can be obtained by adding a further resistor to ground as shown in Figure 3.40.

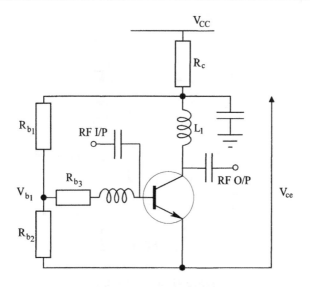

Figure 3.40 RF bias circuit

For this circuit; i_b = 0.1mA, I_c = 5mA, V_{ce} = 5V as before. Now let V_{b1} = 1.7V, then R_{b3} = 10K. Let I_{Rb2} = 1.7mA then R_{b2} = 1kΩ.

$$R_{b1} = \frac{3.3V}{(1.7\text{mA} + i_b)} = 1.83\text{k}\Omega \qquad (3.193)$$

then:

$$R_c = 1.47\text{k}\Omega \qquad (3.194)$$

It is now quite common to use an active bias circuit which uses a pnp transistor (or an op amp) to sense the voltage across the collector resistor and apply the correct bias as shown in Figure 3.41. Here R_1 and R_2 determine V_{ce} and R_3 sets the current.

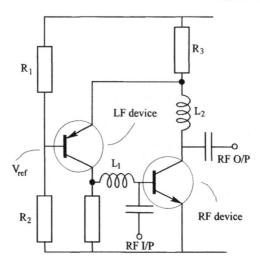

Figure 3.41 Active Bias Circuit

3.9.2   GaAs MESFET biasing

GaAs MESFETs have an I_D vs V_{GS} characteristic shown in Figure 3.42 and it can be seen that the current at zero gate bias (I_{DSS}) can be rather high.

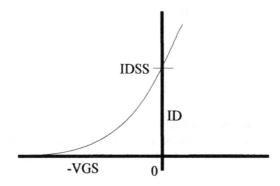

Figure 3.42 I_D vs V_{GS} characteristic of a GaAs MESFET

Bias circuits therefore usually have to be designed to produce a reverse bias on the gate. They also have to be designed to ensure that the power-up characteristics ensure that large peak currents do not flow. In this case a reverse bias should be applied to the gate before the drain source voltage is applied. A typical active bias circuit is shown in Figure 3.43. It is also possible to obtain integrated circuits which contain a switched mode circuit to generate the negative voltage. Care should be taken with these circuits as they often produce sidebands on the carrier frequency at the switching frequency.

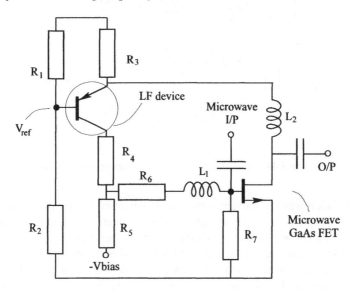

Figure 3.43 A typical GaAs FET biasing circuit

3.10   Measurements and Error Correction

3.10.1   Network Analyser

Most measurements above 5MHz are made using an S parameter network analyser. This type of analyser measures the magnitude and phase of the incident input wave and the magnitude and phase of the transmitted and reflected wave to obtain S_{21} and S_{11}. Switches are then used to turn the device around to measure the reverse transmitted wave and output reflection coefficient S_{12} and S_{22} respectively. A block diagram of a typical network analyser is shown in Figure 3.44.

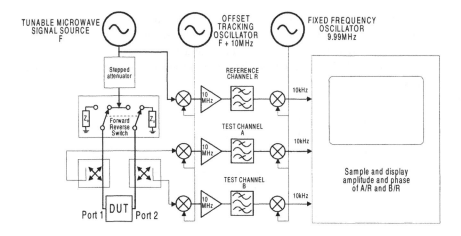

Figure 3.44 Three channel superheterodyne network analyser

An RF/microwave reference source is swept over the frequency range and directional couplers are used to sample the incident transmitted and reflected waves. The signals are measured in a three channel dual superhet receiver, and for this example assume that the first IF is 10 MHz and a second IF is, say, 10kHz. A sample of the incident wave is used as the reference channel R to enable accurate and stable amplitude and phase measurements to be made. The first local oscillator, LO, consists of a second signal source operating 10 MHz above the reference signal source. This can be achieved using phase locked techniques. Some analysers simplify this first LO by using a lower frequency swept signal source which is multiplied up to provide the first LO, where the multiplication ratio changes depending on the band of frequencies covered. All three paths experience gain and further downconversion to an IF of, say, 10kHz. All three signals are then sampled and the ratios A/R and B/R plotted in a variety of formats. Modern analysers typically use synthesised signal sources.

3.10.2   Test Jig

The analyser should be connected to the device via a test jig and the effect of the interconnecting leads removed. It is important to note that the impedance varies along the transmission line and therefore a reference point is required. If the device is active, DC bias is also required. A typical test jig is shown in Figure 3.45 where A and B represent the input and output reference points. The bias inductors (bias tee) provide the DC supply and should present a high impedance at RF frequencies. In-line capacitors are used to provide a DC break to external circuits and should present a low impedance at the operating frequencies.

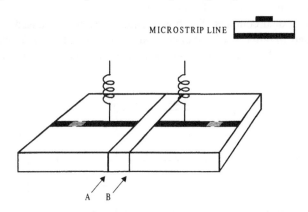

Figure 3.45 S parameter test jig

3.10.3   Calibration and Error Correction

To obtain accurate readings the jig and or interconnecting leads should be calibrated at the reference points A and B. This is usually achieved for each port by sequentially placing a set of loads at the reference point. It will be shown that by using three loads it is possible to deduce the S parameters for the interconnecting cable and the test jig up to the reference point. This therefore enables the effects of the interconnecting paths to be removed.

At RF the calibration loads are typically an O/C, S/C and 50Ω Load.

1.   O/C: This could be as seen. Note that the O/C is often poorly defined at higher frequencies and on commercial connector based calibration kits the frequency dependence of the open circuit is included.

2.   S/C: The transmission line is shorted to ground at A and B.

3.   LOAD: Connect a high performance 50Ω load between ground and A and ground and B.

4.   THROUGH: Remove centre section or use shortened jig.

5.   DEVICE CHANGE: Use different centre section.

Note that measurement of a perfect load on a network analyser without calibration will show ripple (upper curve) as illustrated in Figure 3.46 and a poor return loss due to multiple reflections at the interfaces of the connectors and imperfect directional couplers. The high frequency ripple is caused by reflections spaced far apart in other words, at either end of the cable. Lower frequency ripples are caused by reflections closer together such as those due to adaptors and connector savers.

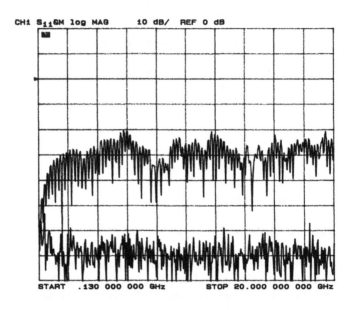

CH1 S₁₁&M log MAG   10 dB/   REF 0 dB

START   .130 000 000 GHz   STOP 20.000 000 000 GHz

Figure 3.46 One port measurements with and without calibration

3.10.4   One Port Error Correction

To obtain accurate measurements it is necessary to correct for the leads and the jig up to the reference points. We can model this interconnecting circuit with an S parameter block as illustrated in Figure 3.47. Here one port error correction will be described in detail. For full measurements both ports and through line error correction are performed. The equations used are the same as those used for the stability analysis.

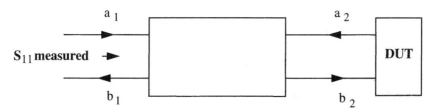

Figure 3.47 The S parameters for the interconnecting block can be deduced from S parameter measurements of the interconnecting block using calibrating loads for the DUT

$$\begin{pmatrix} b_1 \\ b_2 \end{pmatrix} = \begin{pmatrix} S_{11} & S_{12} \\ S_{21} & S_{22} \end{pmatrix} \begin{pmatrix} a_1 \\ a_2 \end{pmatrix}$$ (3.195)

therefore:

$$b_1 = S_{11}a_1 + S_{12}a_2 \text{ and } b_2 = S_{21}a_1 + S_{22}a_2$$ (3.196)

Therefore S_{11} measured is:

$$S_{11M} = \frac{b_1}{a_1} = S_{11} + S_{12}\frac{a_2}{a_1} \text{ as } a_1 = \frac{b_2 - S_{22}a_2}{S_{21}}$$ (3.197)

$$S_{11M} = S_{11} + S_{12}S_{21}\left(\frac{a_2}{b_2 - S_{22}a_2} \right) \text{ (divide by b}_2)$$ (3.198)

$$S_{11M} = S_{11} + S_{12}S_{21}\left(\frac{\Gamma_{DUT}}{1 - S_{22}\Gamma_{DUT}} \right)$$ (3.199)

Therefore placing a good load, an O/C and a S/C in place of the DUT enables the S parameters for the interconnecting block to be deduced. Putting a good load in the place of the DUT causes $\Gamma_{DUT} = 0$. Therefore $S_{11M} = S_{11}$.

Put a S/C in place of the DUT then:

$$\Gamma_{DUT} = -1$$ (3.200)

then:

$$S_{11M} = S_{11} + S_{12}S_{21}\left(\frac{(-1)}{1-S_{22}(-1)}\right)$$

(3.201)

Put an O/C in place of the DUT:

$$\Gamma_{DUT} = +1$$

(3.202)

$$S_{11M} = S_{11} + S_{12}S_{21}\left(\frac{(1)}{1-S_{22}(1)}\right)$$

(3.203)

As $S_{12} = S_{21}$ for passive networks, the S parameters for the interconnecting block can be deduced.

For completeness the full calculations for one port error correction are now given: Let the S parameters of the interconnecting block be:

$$E_{11} = S_{11}$$

(3.204)

$$E_{22} = S_{22}$$

(3.205)

$$E_{21} = E_{12} = S_{21} = S_{12}$$

(3.206)

Note that $S_{21} = S_{12}$ for passive networks and remember when the DUT is a good load, $E_{11} = S_{11}$ measured.

For an O/C termination:

$$\Gamma_{O/C} = E_{11} + \frac{E_{21}^{\;2}}{1-E_{22}}$$

(3.207)

$$\Gamma_{O/C} - E_{11} = -\frac{E_{21}^{\;2}}{1-E_{22}}$$

(3.208)

and for a S/C termination:

$$\Gamma_{S/C} - E_{11} = \frac{E_{21}^{\,2}}{1+E_{22}} \tag{3.209}$$

Dividing (3.208) by (3.209)

$$\frac{\Gamma_{O/C} - E_{11}}{\Gamma_{S/C} - E_{11}} = -\frac{1+E_{22}}{1-E_{22}} = K \tag{3.210}$$

Therefore:

$$-1 - E_{22} = K - KE_{22} \tag{3.211}$$

$$E_{22} = \frac{K+1}{K-1} \tag{3.212}$$

Put (3.212) in (3.208) gives:

$$E_{21}^{\,2} = \left(\Gamma_{O/C} - E_{11}\right)\left(1 - \frac{K+1}{K-1}\right) = \left(\Gamma_{O/C} - E_{11}\right)\left(\frac{K-1}{K-1} - \frac{K+1}{K-1}\right) \tag{3.213}$$

$$E_{21}^{\,2} = 2\frac{\Gamma_{O/C} - E_{11}}{1-K} \tag{3.214}$$

$$E_{21} = \sqrt{2}\left(\frac{\Gamma_{O/C} - E_{11}}{1-K}\right)^{1/2} \tag{3.215}$$

3.11 References and Bibliography

1. Guillermo Gonzalez, *Microwave Transistor Amplifiers. Analysis and Design*, Prentice Hall, 1984.
2. W.H. Haywood, *Introduction to Radio Frequency Design*, Prentice Hall, 1982.
3. P.J. Fish, *Electronic Noise and Low Noise Design*, Macmillan, New Electronics Series, 1993.

4. *RF Wideband Transistors*, Product Selection 2000 Discrete Semiconductors CD (First Edition) Release 01-2000.

5. Chris Bowick, *RF Circuit Design, SAMS, Division of Macmillan*, 1987.

6. H.L. Krauss, C.W. Bostian and F.H. Raab, *Solid State Radio Engineering*, Wiley, 1980.

7. Peter C.L. Yip, *High Frequency Circuit Design and Measurements*, Chapman and Hall, 1990.

8. R.S. Carson, *High Frequency Amplifiers*, Wiley, 1982.

9. *S Parameter Design*, Application Note 154, Hewlett Packard.

10. "Error Models for Systems Measurement", *Microwave Journal*, May, 1973.

11. H.A. Haus (Chairman), *"Representation of Noise in Linear Two Ports"*, IRE subcommittee 7.9 on Noise, Proceedings of the IEEE, January 1960.

12. Philip H. Smith, *Electronic Applications of the Smith Chart*, Noble Publishing 1995.

13. J.G. Linvill and J.F. Gibbons, *Transistors and Active Circuits*, McGraw-Hill, New York, 1961.

14. A.P. Stern "Stability and Power Gain of Tuned Transistor Amplifiers", *Proceedings of the IRE*, **45**, 3, pp.335-343, March 1957.

4

Low Noise Oscillators

4.1   Introduction

The oscillator in communication and measurement systems, be they radio, coaxial cable, microwave, satellite, radar or optical fibre, defines the reference signal onto which modulation is coded and later demodulated. The flicker and phase noise in such oscillators are central in setting the ultimate systems performance limits of modern communications, radar and timing systems. These oscillators are therefore required to be of the highest quality for the particular application as they provide the reference for data modulation and demodulation.

The chapter describes to a large extent a linear theory for low noise oscillators and shows which parameters explicitly affect the noise performance. From these analyses equations are produced which accurately describe oscillator performance usually to within 0 to 2dB of the theory. It will show that there are optimum coupling coefficients between the resonator and the amplifier to obtain low noise and that this optimum is dependent on the definitions of the oscillator parameters. The factors covered are:

1.   The noise figure (and also source impedance seen by the amplifier).

2.   The unloaded Q, the resonator coupling coefficient and hence Q_L/Q_0 and closed loop gain.

3.   The effect of coupling power out of the oscillator.

4.   The loop amplifier input and output impedances and definitions of power in the oscillator.

5.   Tuning effects including the varactor Q and loss resistance, and the coupling coefficient of the varactor.

6.   The open loop phase shift error prior to loop closure.

Optimisation of parameters using a linear analytical theory is of course much easier than non-linear theories.

The chapter then includes eight design examples which use inductor/capacitor, surface acoustic wave (SAW), transmission line, helical and dielectric resonators at 100MHz, 262MHz, 900MHz, 1800MHz and 7.6GHz. These oscillator designs show very close correlation with the theory usually within 2dB of the predicted minimum. The Chapter also includes a detailed design example.

The chapter then goes on to describe the four techniques currently available for flicker noise measurement and reduction including the latest techniques developed by the author's research group in September 2000, in which a feedforward amplifier is used to suppress the flicker noise in a microwave GaAs based oscillator by 20dB. The theory in this chapter accurately describes the noise performance of this oscillator within the thermal noise regime to within ½ to 1dB of the predicted minimum.

A brief introduction to a method for breaking the loop at any point, thus enabling non-linear computer aided analysis of oscillating (autonomous) systems is described. This enables prediction of the biasing, output power and harmonic spectrum.

4.2   Oscillator Noise Theories

The model chosen to analyse an oscillator is extremely important. It should be simple, to enable physical insight, and at the same time include all the important parameters. For this reason both equivalent circuit and block diagram models are presented here. Each model can produce different results as well as improving the understanding of the basic model. The analysis will start with an equivalent circuit model, which allows easy analysis and is a general extension of the model originally used by the author to design high efficiency oscillators [2]. This was an extension of the work of Parker who was the first to discuss noise minima in oscillators in a paper on surface acoustic wave oscillators [1]. Two definitions of power are used which produce different optima. These are P_{RF} (the power dissipated in the source, load and resonator loss resistance) and the power available at the output P_{AVO} which is the maximum power available from the output of the amplifier which would be produced into a matched load. It is important to consider both definitions. The use of P_{AVO} suggests further optima (that the source and load impedance should be the same), which is incorrect and does not enable the design of highly power efficient low phase noise designs which inherently require low (zero) output impedance.

The general equivalent circuit model is then modified to model a high efficiency oscillator by allowing the output impedance to drop to zero. This has recently been used to design highly efficient low noise oscillators at L band [6] which demonstrate very close correlation with the theory.

4.3　　Equivalent Circuit Model

The first model is shown in Figure 4.1 and consists of an amplifier with two inputs with equal input impedance, one to model noise (V_{IN2}) and one as part of the feedback resonator (V_{IN1}). In a practical circuit the amplifier would have a single input, but the two inputs are used here to enable the noise input and feedback path to be modelled separately. The signals on the two inputs are therefore added together. The amplifier model also has an output impedance (R_{OUT}).

The feedback resonator is modelled as a series inductor capacitor circuit with an equivalent loss resistance R_{LOSS} which defines the unloaded Q (Q_0) of the resonator as $\omega L/R_{LOSS}$. Any impedance transformations are incorporated into the model by modifying the LCR ratios.

The operation of the oscillator can best be understood by injecting white noise at input V_{IN2} and calculating the transfer function while incorporating the usual boundary condition of $G\beta_0 = 1$ where G is the limited gain of the amplifier when the loop is closed and β_0 is the feedback coefficient at resonance.

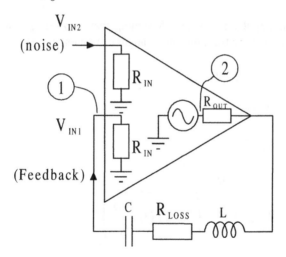

Figure 4.1 Equivalent circuit model of oscillator

The noise voltage V_{IN2} is added at the input of the amplifier and is dependent on the input impedance of the amplifier, the source resistance presented to the input of the amplifier and the noise figure of the amplifier. In this analysis, the noise figure under operating conditions, which takes into account all these parameters, is defined as F.

The circuit configuration is very similar to an operational amplifier feedback circuit and therefore the voltage transfer characteristic can be derived in a similar way. Then:

$$V_{OUT} = G(V_{IN2} + V_{IN1}) = G(V_{IN2} + \beta V_{OUT}) \qquad (4.1)$$

where G is the voltage gain of the amplifier between nodes 2 and 1, β is the voltage feedback coefficient between nodes 1 and 2 and V_{IN2} is the input noise voltage. The voltage transfer characteristic is therefore:

$$\frac{V_{OUT}}{V_{IN2}} = \frac{G}{1 - (\beta G)} \qquad (4.2)$$

By considering the feedback element between nodes 1 and 2, the feedback coefficient is derived as:

$$\beta = \frac{R_{IN}}{R_{LOSS} + R_{IN} + j(\omega L - 1/\omega C)} \qquad (4.3)$$

Where ω is the angular frequency. Assuming that: $\Delta\omega \ll \omega_0$ (where $\Delta\omega$ is the offset angular frequency from the centre angular frequency ω_0) then as:

$$(\omega L - 1/\omega C) = \pm 2\Delta\omega L \qquad (4.4)$$

and as the loaded Q is:

$$Q_L = \omega_0 L / (R_{OUT} + R_{LOSS} + R_{IN}) \qquad (4.5)$$

$$\beta = \frac{R_{IN}}{(R_{OUT} + R_{LOSS} + R_{IN})\left(1 \pm 2j Q_L \dfrac{\Delta\omega}{\omega_0}\right)} \qquad (4.4)$$

The unloaded Q is:

$$Q_0 = \omega_0 L / R_{LOSS} \qquad (4.6)$$

then:

$$Q_L / Q_0 = R_{LOSS} / (R_{OUT} + R_{OUT} + R_{IN}) \qquad (4.7)$$

As:

$$\left(1 - Q_L/Q_0\right) = \left(R_{OUT} + R_{IN}\right)/\left(R_{OUT} + R_{LOSS} + R_{IN}\right) \qquad (4.8)$$

then the feedback coefficient at resonance, β_0, between nodes 1 and 2, is:

$$\beta_0 = R_{IN}/\left(R_{OUT} + R_{LOSS} + R_{IN}\right) = \left(1 - Q_L/Q_0\right)\left(\frac{R_{IN}}{R_{IN} + R_{OUT}}\right) \qquad (4.9)$$

Therefore the resonator response is:

$$\beta = \left(\frac{R_{IN}}{R_{IN} + R_{OUT}}\right)\left(1 - \frac{Q_L}{Q_0}\right)\frac{1}{\left(1 \pm 2jQ_L\dfrac{df}{f_0}\right)} \qquad (4.10)$$

where f_0 is now the centre frequency and Δf is the offset frequency from the carrier now in Hertz. In fact if $R_{OUT} = R_{IN}$ then the scattering parameter S_{21} can be calculated as 2β therefore:

$$S_{21} = \left(1 - \frac{Q_L}{Q_0}\right)\frac{1}{\left(1 \pm 2jQ_L\dfrac{df}{f_0}\right)} \qquad (4.11)$$

This is a general equation which describes the variation of insertion loss (S_{21}) of most resonators with selectivity and hence loaded (Q_L) and unloaded Q (Q_0). The first term shows how the insertion loss varies with selectivity at the center frequency and that maximum insertion loss occurs when Q_L tends to Q_0 at which point the insertion loss tends to infinity. This is illustrated in Figure 4.2 and can be used to obtain the unloaded Q_0 of resonators by extrapolating measurement points via a straight line to the intercept Q_0.

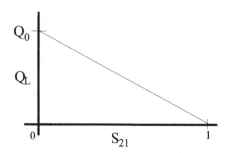

Figure 4.2 S_{21} vs Q_L

The second term, in equation (4.11) describes the frequency response of the resonator for $\Delta f/f_0 \ll 1$. The voltage transfer characteristic of the closed loop, where V_{OUT} is the output voltage of the amplifier, is therefore:

$$\frac{V_{OUT}}{V_{IN2}} = \frac{G}{1 - \dfrac{G\left(1 - Q_L/Q_0\right)\left(\dfrac{R_{IN}}{R_{OUT} + R_{IN}}\right)}{\left(1 \pm 2jQ_L \dfrac{\Delta f}{f_0}\right)}} \tag{4.12}$$

At resonance Δf is zero and V_{OUT}/V_{IN2} is very large. The output voltage is defined by the maximum swing capability of the amplifier and the input voltage is noise. The denominator of eqn. (4.12) is approximately zero, therefore $G\beta_0 = 1$ and:

$$G = \frac{1}{\left(1 - Q_L/Q_0\right)\left(\dfrac{R_{IN}}{R_{OUT} + R_{IN}}\right)} \tag{4.13}$$

This is effectively saying that at resonance the amplifier gain is equal to the insertion loss. The gain of the amplifier is now fixed by the operating conditions. Therefore:

$$\tag{4.14}$$

$$\frac{V_{OUT}}{V_{IN2}} = \frac{G}{1 - \dfrac{1}{1 \pm \left(2jQ_L \dfrac{df}{f_0}\right)}} = \frac{1}{\left(1 - Q_L/Q_0\right)\left(\dfrac{R_{IN}}{R_{OUT} + R_{IN}}\right)\left(1 - \dfrac{1}{1 \pm \left(2jQ_L \dfrac{df}{f_0}\right)}\right)}$$

In oscillators it is possible to make one further approximation if we wish to consider just the 'skirts' of the sideband noise as these occur within the 3dB bandwidth of the resonator. The Q multiplication process causes the noise to fall to the noise floor within the 3dB bandwidth of the resonator. As the noise of interest therefore occurs within the boundaries of $Q_L \Delta f/f_0 \ll 1$ equation (4.14) simplifies to:

$$\frac{V_{OUT}}{V_{IN2}} = \frac{G}{\pm 2 j Q_L \dfrac{\Delta f}{f_o}} = \frac{1}{\left(1 - Q_L/Q_0\right)\left(\dfrac{R_{IN}}{R_{OUT}+R_{IN}}\right)\left(\pm 2 j Q_L \dfrac{\Delta f}{f_0}\right)} \qquad (4.15)$$

Note therefore that the gain has been incorporated into the equation in terms of Q_L/Q_0 as the gain is set by the insertion loss of the resonator.

It should be noted, however, that this equation does not apply very close to carrier where V_{out} approaches and exceeds the peak voltage swing of the amplifier. As:

$$V_{IN} = \sqrt{4kTBR_{IN}} \qquad (4.16)$$

which is typically 10^{-9} in a 1Hz bandwidth and V_{OUT} is typically 1 volt, G is typically 2, then for this criteria to apply $Q_L \Delta f/f_0 \gg 10^{-9}$. For a Q_L of 50, centre frequency of $f_0 = 10^9$ Hz, errors only start to occur at frequency offsets closer than 1 Hz to carrier. In fact this effect is slightly worse than a simple calculation would suggest as PM is a non-linear form of modulation. It can only be regarded as linear for phase deviations much less than 0.1 rad.

As the sideband noise in oscillators is usually quoted as power not voltage it is necessary to define output power. It is also necessary to decide where the limiting occurs in the amplifier. In this instance limiting is assumed to occur at the output of the amplifier, as this is the point where the maximum power is defined by the power supply. In other words, the maximum voltage swing is limited by the power supply.

Noise in oscillators is usually quoted in terms of a ratio. This ratio L_{FM} is the ratio of the noise in a 1Hz bandwidth at an offset Δf over the total oscillator power as shown in Figure 4.3.

To investigate the ratio of the noise power in a 1 Hz sideband to the total output power, the voltage transfer characteristic can now be converted to a characteristic which is proportionate to power. This was achieved by investigating the square of the output voltage at the offset frequency and the square of the total output voltage. Only the power dissipated in the oscillating system and not the power dissipated in the load is included.

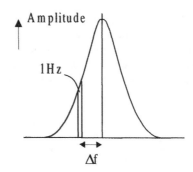

Figure 4.3 Phase noise variation with offset

The input noise power in a 1Hz bandwidth is FkT (k is Boltzman's constant and T is the operating temperature) where kT is the noise power that would have been available at the input had the source impedance been equal to the input impedance (R_{IN}). F is the operating noise figure which includes the amplifier parameters under the oscillating operating conditions. This includes such parameters as source impedance. The dependence of F with source impedance is discussed later in the chapter. The square of the input voltage is therefore $FkTR_{IN}$.

It should be noted that the noise voltage generated by the series loss resistor in the tuned circuit was taken into account by the noise figure of the amplifier. The important noise was within the bandwidth of the tuned circuit allowing the tuned circuit to be represented as a resistor over most of the performance close to carrier. In fact the sideband noise power of the oscillator reaches the background level of noise around the 3dB point of the resonator.

The noise power is usually measured in a 1Hz bandwidth. The square of the output voltage in a 1Hz bandwidth at a frequency offset Δf is:

$$(V_{OUT}\,\Delta f)^2 = \frac{FkTR_{IN}}{4(Q_L)^2\,(R_{IN}/(R_{OUT}+R_{IN}))^2\,(1-Q_L/Q_0)^2}\left(\frac{f_0}{\Delta f}\right)^2 \quad (4.17)$$

Note that parameters such as Q_0 are fixed by the type of resonator. However, Q_L/Q_0 can be varied by adjusting the insertion loss (and hence coupling coefficient) of the resonator. The denominator of equation (4.17) is therefore separated into constants such as Q_0 and variables in terms of Q_L/Q_0. Note that the insertion loss of the resonator also sets the closed loop gain of the amplifier. This equation therefore also includes the effect of the closed loop amplifier gain on the noise performance.

Equation (4.17) can therefore be rewritten in a way which separates the constants and variables as:

$$(V_{OUT} \Delta f)^2 = \frac{FkTR_{IN}}{4(Q_0)^2 (Q_L/Q_0)^2 (R_{IN}/(R_{OUT} + R_{IN}))^2 (1 - Q_L/Q_0)^2} \left(\frac{f_0}{\Delta f}\right)^2 \quad (4.18)$$

As this theory is a linear theory, the sideband noise is effectively amplified narrow band noise. To represent this as an ideal carrier plus sideband noise, the signal can be thought of as a carrier with a small perturbation rotating around it as shown in Figure 4.4.

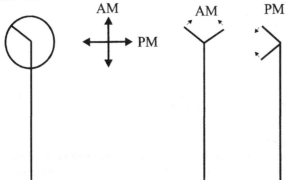

Figure 4.4 Representation of signal with AM and PM components

Note that there are two vectors rotating in opposite directions, one for the upper and one for the lower sideband. The sum of these vectors can be thought of as containing both amplitude modulation (AM) and phase modulation (PM). The component along the axis of the carrier vector being AM noise and the component orthogonal to the carrier vector being phase noise. PM can be thought of as a linear modulation as long as the phase deviation is considerably less than 0.1 rad.

Equation (4.18) accurately describes the noise performance of an oscillator which uses automatic gain control (AGC) to define the output power. However, the theory would only describe the noise performance at offsets greater than the AGC loop bandwidth.

Although linear, this theory can incorporate the non-linearities, i.e. limiting in the amplifier, by modifying the absolute value of the noise. If the output signal amplitude is limited with a 'hard' limiter, the AM component would disappear and the phase component would be half of the total value shown in equation (4.18). This is because the input noise is effectively halved. This assumes that the limiting does not cause extra components due to mixing. Limiting also introduces a form of coherence between the upper and the lower sideband which has been defined by Robins [5] as conformability. The square of the output voltage is therefore:

$$(V_{OUT}\Delta f)^2 = \frac{FkTR_{IN}}{8(Q_0)^2 (Q_L/Q_0)^2 (R_{IN}/(R_{OUT}+R_{IN}))^2 (1-Q_L/Q_0)^2}\left(\frac{f_0}{\Delta f}\right)^2 \quad (4.19)$$

The output noise performance is usually defined as a ratio of the sideband noise power to the total output power. If the total output voltage is $V_{OUTMAXRMS}$, the ratio of sideband phase noise, in a 1Hz bandwidth, to total output will be L_{FM}, therefore:

$$L_{FM} = \frac{(V_{OUT}\,\Delta f)^2}{(V_{OUT\,MAX\,RMS})^2} \quad (4.20)$$

$$L_{FM} = \frac{FkTR_{IN}}{8(Q_0)^2 (Q_L/Q_0)^2 (1-Q_L/Q_0)^2 (R_{IN}/(R_{OUT}+R_{IN}))^2 (V_{OUT\,MAX\,RMS})^2}\left(\frac{f_0}{\Delta f}\right)^2 \quad (4.21)$$

When the total RF feedback power, P_{RF}, is defined as the power in the oscillating system, excluding the losses in the amplifier, and most of the power is assumed to be close to carrier, then P_{RF} is limited by the maximum voltage swing at the output of the amplifier and the value of $R_{OUT} + R_{LOSS} + R_{IN}$.

$$P_{RF} = \frac{(V_{OUT\,MAX\,RMS})^2}{R_{OUT}+R_{LOSS}+R_{IN}} \quad (4.22)$$

Equation (4.21) becomes:

$$L_{FM} = \frac{FkT(R_{OUT}+R_{IN})^2}{8(Q_0)^2 (Q_L/Q_0)^2 R_{IN} (1-Q_L/Q_0)^2 P_{RF}(R_{OUT}+R_{LOSS}+R_{IN})}\left(\frac{f_0}{\Delta f}\right)^2 \quad (4.23)$$

As:

$$\frac{R_{OUT}+R_{IN}}{R_{OUT}+R_{LOSS}+R_{IN}} = (1-Q_L/Q_0) \quad (4.24)$$

The ratio of sideband noise in a 1Hz bandwidth at offset Δf to the total power is therefore:

$$L_{FM} = \frac{FkT}{8(Q_0)^2 (Q_L/Q_0)^2 (1-Q_L/Q_0)P_{RF}}\left(\frac{R_{OUT}+R_{IN}}{R_{IN}}\right)\left(\frac{f_0}{\Delta f}\right)^2 \quad (4.25)$$

If R_{OUT} is zero as in the case of a high efficiency oscillator, this equation simplifies to:

$$L_{FM} = \frac{FkT}{8(Q_0)^2 (Q_L/Q_0)^2 (1 - Q_L/Q_0)P_{RF}} \left(\frac{f_0}{\Delta f}\right)^2 \qquad (4.26)$$

Note of course that F may well vary with source impedance which will vary as Q_L/Q_0 varies.

Equation (4.20) illustrates the fact that the output impedance serves no useful purpose other than to dissipate power. If R_{OUT} is allowed to equal R_{IN} as might well be the case in many RF and microwave amplifiers then equation (4.25) simplifies to:

$$L_{FM} = \frac{FkT}{4(Q_0)^2 (Q_L/Q_0)^2 (1 - Q_L/Q_0)P_{RF}} \left(\frac{f_0}{\Delta f}\right)^2 \qquad (4.27)$$

It should be noted that P_{RF} is the total power in the system excluding the losses in the amplifier, from which: P_{RF} = (DC input power to the system) × efficiency.

When the power in the oscillator is defined as the power available at the output of the amplifier P_{AVO} then:

$$P_{AVO} = \frac{(V_{OUT\,MAX\,RMS})^2}{4R_{OUT}} \qquad (4.28)$$

Equation (4.21) then becomes:

$$L_{FM} = \frac{FkT \; R_{IN}}{8(Q_0)^2 (Q_L/Q_0)^2 (R_{IN}/(R_{OUT} + R_{IN}))^2 (1 - Q_L/Q_0)^2 P_{AVO}(4R_{OUT})} \left(\frac{f_0}{\Delta f}\right)^2 \; (4.29)$$

which can be re-arranged as:

$$L_{FM} = \frac{FkT}{32(Q_0)^2 (Q_L/Q_0)^2 (1 - Q_L/Q_0)^2 P_{AVO}} \left(\frac{(R_{OUT} + R_{IN})^2}{R_{OUT} \cdot R_{IN}}\right) \left(\frac{f_0}{\Delta f}\right)^2 \qquad (4.30)$$

The term:

$$\left(\frac{\left(R_{OUT} + R_{IN} \right)^2}{R_{OUT} \cdot R_{IN}} \right) \tag{4.31}$$

can be shown to be minimum when $R_{OUT} = R_{IN}$ and is then equal to four. However this is only because the definition of power is the power available at the output of the amplifier, P_{AVO}. As R_{OUT} reduces P_{AVO} gets larger and the noise performance gets worse, however P_{AVO} then relates less and less to the actual power in the oscillator. If $R_{OUT} = R_{IN}$:

$$L_{FM} = \frac{FkT}{8 \left(Q_0 \right)^2 \left(Q_L / Q_0 \right)^2 \left(1 - Q_L / Q_0 \right)^2 P_{AVO}} \left(\frac{f_0}{\Delta f} \right)^2 \tag{4.32}$$

A general equation can then be written which describes all three cases:

$$L_{FM} = A \cdot \frac{FkT}{8 \left(Q_0 \right)^2 \left(Q_L / Q_0 \right)^2 \left(1 - Q_L / Q_0 \right)^N P} \left(\frac{f_0}{\Delta f} \right)^2 \tag{4.33}$$

where:

1. $N = 1$ and $A = 1$ if P is defined as P_{RF} and $R_{OUT} =$ zero.

2. $N = 1$ and $A = 2$ if P is defined as P_{RF} and $R_{OUT} = R_{IN}$.

3. $N = 2$ and $A = 1$ if P is defined as P_{AVO} and $R_{OUT} = R_{IN}$.

This equation describes the noise performance within the 3dB bandwidth of the resonator which rolls off as $(1/\Delta f)^2$, as predicted by Leeson [43] and Cutler and Searle [44] in their early models, but a number of further parameters are also included in this new equation.

Equation (4.33) shows that L_{FM} is inversely proportional to P_{RF} and that a larger ratio is thus obtained for higher feedback power. This is because the absolute value of the sideband power, at a given offset, does not vary with the total feedback power. This is illustrated in Figure 4.5 where it is seen that the total power is effectively increased by increasing the power very close to carrier.

The noise performance outside the 3dB bandwidth is just the product of the closed loop gain, noise figure and the thermal noise if the output is taken at the output of the amplifier (although this can be reduced by taking the output after the resonator). The spectrally flat part of the spectrum is not included in equation (4.33) as the aim, in this chapter, is to reduce the $(1/\Delta f)^2$ spectrum to a minimum.

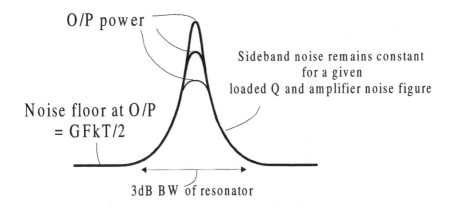

O/P power

Sideband noise remains constant
for a given
loaded Q and amplifier noise figure

Noise floor at O/P
= GFkT/2

3dB BW of resonator

Figure 4.5   Noise spectrum variation with RF power

4.4   The Effect of the Load

Note that the models used so far have ignored the effect of the load. If the output impedance of the amplifier was zero, the load would have no effect. If there is a finite output impedance then the load will of course have an effect which has not been included so far. However, the load can most easily be incorporated as a coupler/attenuator at the output of the amplifier which causes a reduction in the maximum open loop gain and an increase in the amplifier noise figure. The closed loop gain, of course, does not change as this is set by the insertion loss of the resonator.

4.5   Optimisation for Minimum Phase Noise

4.5.1   *Models Using Feedback Power Dissipated in the Source, Resonator Loss and Input Resistance*

If the power is defined as P_{RF} then the following equation was derived

$$L_{FM} = A \cdot \frac{FkT}{8 \, (Q_0)^2 \, (Q_L/Q_0)^2 \, (1 - Q_L/Q_0)P} \left(\frac{f_0}{\Delta f} \right)^2 \qquad (4.34)$$

where:

1. $A = 1$ if P is defined as P_{RF} and R_{OUT} = zero.
2. $A = 2$ if P is defined as P_{RF} and $R_{OUT} = R_{IN}$.

This equation should now be differentiated in terms of Q_L/Q_0 to determine where there is a minimum. At this stage we will assume that the ratio of R_{OUT}/R_{IN} is either zero or fixed to a finite value. Therefore the phase noise equation is minimum when:

$$\frac{dL_{FM}}{d(Q_L/Q_0)} = 0 \qquad (4.35)$$

Minimum noise therefore occurs when $Q_L/Q_0 = 2/3$. To satisfy $Q_L/Q_0 = 2/3$, the voltage insertion loss of the resonator between nodes 2 and 1 is 1/3 which sets the amplifier voltage gain, between nodes 1 and 2, to 3.

It is extremely important to use the correct definition of power (P), as this affects the values of the parameters required to obtain optimum noise performance.

4.5.2   Models Using Power at the Input as the Limited Power

If the power is defined as the power at the input of the amplifier (P_I) then the gain/insertion loss will disappear from the equation to produce:

$$L_{FM} = \frac{FkT}{8Q_L^2 P_I} \left(\frac{f_0}{\Delta f}\right)^2 \qquad (4.36)$$

At first glance it would appear that minimum noise occurs when Q_L is made large and hence tends to Q_0. However this would increase the insertion loss requiring the amplifier gain and output power both to tend to infinity

4.5.3   Models Using Power Available at the Output as the Limited Power

If the power is now defined as the power available at the output of R_{OUT} i.e. $P_{AVO} = (V^2/R_{OUT})$ as shown in the equivalent circuit model (Figure 4.1), then this produces the same answer as that produced by the block diagram oscillator model shown in Figure 4.6 where V is the voltage before the output resistance R_{out} at node 2.

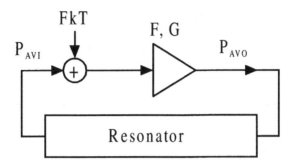

Figure 4.6 Block Diagram Model

The following equation was derived:

$$L_{FM} = \frac{FkT}{32(Q_0)^2 (Q_L/Q_0)^2 (1 - Q_L/Q_0)^2 P_{AVO}} \left(\frac{(R_{OUT} + R_{IN})^2}{R_{OUT} \cdot R_{IN}} \right) \left(\frac{f_0}{\Delta f} \right)^2 \quad (4.30)$$

The term:

$$\left(\frac{(R_{OUT} + R_{IN})^2}{R_{OUT} \cdot R_{IN}} \right) \qquad (4.31)$$

can be shown to be minimum when $R_{OUT} = R_{IN}$ which then sets this term equal to four. However, this is only because the definition of power is now the power available at the output of the amplifier which is not directly linked to the power in the oscillator. If $R_{OUT} = R_{IN}$ then equation 3 simplifies to:

$$L_{FM} = \frac{FkT}{8(Q_0)^2 (Q_L/Q_0)^2 (1 - Q_L/Q_0)^2 P_{AVO}} \left(\frac{f_0}{\Delta f} \right)^2 \qquad (4.32)$$

The minimum of equations (4.30) and (4.32) occurs when $Q_L/Q_0 = 1/2$. It should be noted that P_{AVO} is constant and not related to Q_L/Q_0. The power available at the output of the amplifier is different from the power dissipated in the oscillator, but by chance is close to it. Parker [1] has shown a similar optimum for SAW oscillators and was the first to mention an optimum ratio of Q_L/Q_0. Moore and Salmon also incorporate this in their paper [7].

Equations (4.32) should be compared with the model in which P_{RF} is limited where the term in the denominator has now changed from $(1 - Q_L/Q_0)$ to $(1 - Q_L/Q_0)^2$.

These results are most easily compared graphically as shown in Figure 4.7. Measurements of noise variation with Q_L/Q_0 have been demonstrated using a low frequency oscillator [2] where the power is defined as P_{RF} and these are also included in Figure 4.7.

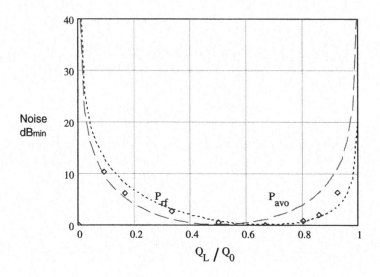

Figure 4.7 Phase Noise vs Q_L/Q_0 for the two different definitions of power

The difference in the noise performance and the optimum operating point predicted by the different definitions of power is small. However, care needs to be taken when using the P_{AVO} definition if it is necessary to know the optimum value of the source and load impedance. For example, if P_{AVO} is fixed it would appear that optimum noise performance would occur when $R_{OUT} = R_{IN}$ because P_{AVO} tends to be very large when R_{OUT} tends to zero. This is not the case when P_{RF} is fixed, as P_{RF} does not require a matched load.

4.5.4   Effect of Source Impedance on Noise Factor

It should also be noted that the noise factor is dependent on the source impedance presented to the amplifier and that this will change the optimum operating point depending on the type of active device used. If the variation of noise performance with source impedance is known, as illustrated in Figure 4.8, then this can be incorporated to slightly shift the optimum value of Q_L/Q_0. Further it is often possible to vary the ratio of the optimum source impedance to input impedance in bipolar transistors using, for example, an emitter inductor as described in chapter 3.

This then enables the input impedance and the optimum source impedance to be chosen separately for minimum noise. This inductor, if small, causes very little change in the noise performance but changes the real part of the input impedance due to the product of the imaginary part of the complex current gain β and the emitter load $j\omega l$.

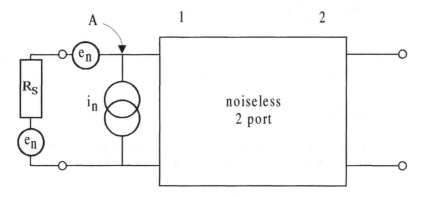

Figure 4.8 Typical noise model for the active device

4.6   Noise Equation Summary

In summary a general equation can then be written which describes all three cases:

$$L_{FM} = A \cdot \frac{FkT}{8\,(Q_0)^2\,(Q_L/Q_0)^2\,(1 - Q_L/Q_0)^N\,P}\left(\frac{f_0}{\Delta f}\right)^2 \qquad (4.33)$$

1. $N = 1$ and $A = 1$ if P is defined as P_{RF} and R_{OUT} = zero.

2. $N = 1$ and $A = 2$ if P is defined as P_{RF} and $R_{OUT} = R_{IN}$.

3. $N = 2$ and $A = 1$ if P is defined as P_{AVO} and $R_{OUT} = R_{IN}$

If the oscillator is operating under optimum operating conditions, then the noise performance incorporating the total RF power (P_{RF}) ($Q_L/Q_0 = 2/3$) simplifies to:

$$L_{FM} = \frac{A \cdot 27 FkT}{32 Q_0^{\,2}\,P_{RF}}\left(\frac{f_0}{\Delta f}\right)^2 \qquad (4.37)$$

where $A = 1$ if R_{OUT} = zero, and $A = 2$ if $R_{OUT} = R_{IN}$.

The noise equation when the power is defined as the power available from the output (P_{AVO}), $R_{OUT} = R_{IN}$ and $Q_L/Q_0 = 1/2$ simplifies to:

$$L_{FM} = \frac{2FkT}{Q_0^2 P_{AVO}} \left(\frac{f_0}{\Delta f} \right)^2 \tag{4.38}$$

The difference is largely explained by the fact that the power available does not include the power dissipated in the output resistor.

4.7   Oscillator Designs

A number of low noise oscillators have been built using the theories for minimum phase noise described in this chapter and these are now described.

4.7.1   Inductor Capacitor Oscillators

A 150 MHz inductor capacitor oscillator [2][3][4] is shown in Figure 4.9. The amplifier operates up to 1Ghz with near 50Ω input and output impedances. The resonator consists of a series tuned LC circuit ($L = 235$nH) with a Q_0 around 300. This sets the series loss resistance of this inductor to be 0.74Ω.

To obtain $Q_L/Q_0 = 1/2$, LC matching networks were added at each end to transform the 50Ω impedances of the amplifier to be $(0.5 \times 0.74)\Omega = 0.37\Omega$. Note the series L of the transformer merges with the L of the tuned circuit. To obtain such large transformation ratios high value capacitors were used and therefore the parasitic inductance of these components should be incorporated. The resonator therefore had an insertion loss of 6 dB and a loaded $Q = 150$. The measured phase noise performance at 1kHz offset was -106.5dBc/Hz. The theory predicts -108dBc/Hz, assuming the transposed flicker noise corner $=1$kHz causing an increase of 3dB above the thermal noise equation.

At 5kHz offset the measured phase noise was -122.3dBc/Hz (theory -125dBc/Hz) and at 10kHz -128.3dBc/Hz (theory -131dBc/Hz).

The following parameters were assumed for the theoretical calculations: $Q_0 = 300$, $P_{AVO} = 1$mW, noise figure = 6dB. The flicker noise corner was measured to be around 1kHz. These measurements are therefore within 3dB of the predicted minimum.

This oscillator is a similar configuration to the Pearce oscillator but the design equations for minimum noise are quite different. A detailed design example illustrating the design process for this type of oscillator is shown at the end of this chapter in section 4.13.

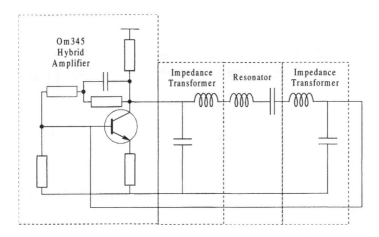

Figure 4.9 Low noise *LC* oscillator

4.7.2   SAW Oscillators

A 262 MHz SAW oscillator using an STC resonator with an unloaded Q of 15,000 was built by Curley and Everard in 1987 [41]. This oscillator was built using low cost components and the noise performance was measured to be better than - 130dBc/Hz at 1kHz, where the flicker noise corner of the measurement was around 1kHz. This noise performance was in fact limited by the measurement system. The oscillator consisted of a resonator with an unloaded Q of 15,000, impedance transforming and phase shift networks and a hybrid amplifier as shown in Figure 4.10. The phase shift networks are designed to ensure that the circuit oscillates on the peak of the amplitude response of the resonator and hence at the maximum in the phase slope ($d\phi/d\omega$). The oscillator will always oscillate at phase shifts of N*360° where N is an integer, but if this is not on the peak of the resonator characteristic, the noise performance will degrade with a $\cos^4\theta$ relationship as discussed later in Section 4.8.4.

Montress, Parker, Loboda and Greer [20] have demonstrated some excellent 500 MHz SAW oscillator designs where they reduced the Flicker noise in the resonators and operated at high power to obtain -140 dBc/Hz at 1kHz offset. The noise performance appeared to be flicker noise limited over the whole offset band.

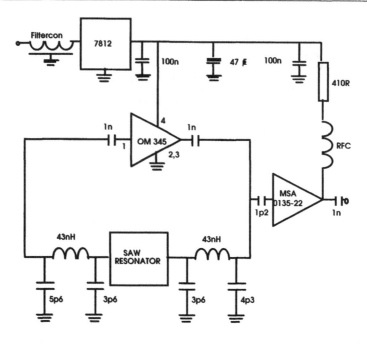

Figure 4.10 Low noise 262 MHz SAW oscillator

4.7.3   Transmission Line Oscillators

Figure 4.11 illustrates a transmission line oscillator [21] [22]. Here the resonator operation is similar to that of an optical Fabry Pérot resonator and the shunt capacitors act as mirrors. The value of the capacitors are adjusted to obtain the correct insertion loss and Q_L/Q_0 calculated from the loss of the transmission line.

The resonator consists of a low-loss transmission line (length L) and two shunt reactances of normalized susceptance jX. If the shunt element is a capacitor of value C then $X = 2\pi f C Z_0$. The value of X should be the effective susceptance of the capacitor as the parasitic series inductance is usually significant. These reactances can also be inductors, an inductor and capacitor, or shunt stubs.

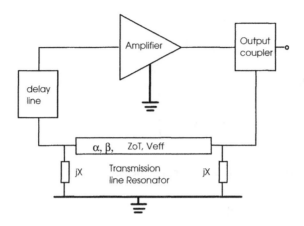

Figure 4.11 Transmission line oscillator

The transmission coefficient of the resonator, S_{21}, can be shown to be:

$$S_{21} = \frac{4z\Gamma}{\left(2+\Phi^2\right)-\Phi^2\Gamma^2}$$

(4.39)

where:

$$\Gamma = \exp\left[-\left(\alpha+j\beta\right)L\right]$$

(4.40)

$$\Phi = z + zX - 1$$

(4.41)

$$z = \frac{Z_T}{Z_0}$$

(4.42)

Z_T is the resonator transmission line impedance, Z_0 is the terminating impedance, α and β are the attenuation coefficient and phase constant of the transmission line respectively. The resonant frequency can be shown to be:

$$f_0 = \frac{V_{EFF}}{2L}\left[1+\left(\frac{1}{\pi}\right)\tan^{-1}\left(\frac{2Xz}{X^2z^2+z^2-1}\right)\right]$$

(4.43)

The insertion loss at resonance is therefore:

$$|S_{21}(0)| = \frac{4z}{4z + 2\alpha L\left[(z-1)^2 + X^2 z^2\right]}$$
(4.44)

If $Q_L \gg \pi$ then:

$$Q_L = \frac{\pi}{4z}|S_{21}(0)|\left((z-1)^2 + X^2 z^2\right)$$
(4.45)

When the phase shift of the resonator is neglected:

$$S_{21}(\Delta f) = \frac{|S_{21}(0)|}{1 + 2jQ_L(\Delta f / f_0)}$$
(4.46)

$$|S_{21}(0)| = 1 - Q_L / Q_0$$
(4.47)

$$Q_0 = \frac{\pi}{2\alpha L}$$
(4.48)

If $Z_T = Z_0$, where Z_T is the resonator line impedance and Z_0 is the terminating impedance and α is the voltage attenuation coefficient of the line β is the phase constant of the line. For small αL (< 0.05) and $\Delta f/f_0 \ll 1$, the following properties can be derived for the first resonant peak (f_0) of the resonator where $\Delta f = f - f_0$ then equation (4.43) simplifies to:

$$f_0 = \left(\frac{V_{eff}}{2L}\right)\left(1 + \left(\frac{1}{\pi}\right)\tan^{-1}\left(\frac{2}{X}\right)\right)$$
(4.49)

Equation (4.45) simplifies to:

$$Q_L = \pi S_{21}(0)\left(\frac{X^2}{4}\right)$$
(4.50)

And:

$$S_{21}(0) = (1 - Q_L/Q_0) = \cfrac{1}{\left(1 + \left(\cfrac{\alpha L}{2}\right)X^2\right)} \qquad (4.51)$$

From these equations it can be seen that the insertion loss and the loaded Q factor of the resonator are interrelated. In fact as the shunt capacitors (assumed to be lossless) are increased the insertion loss approaches infinity and Q_L increases to a limiting value of $\pi/2\alpha L$ which we have defined as Q_0. It is interesting to note that when $S_{21} = 1/2$, $Q_L = Q_0/2$.

4.7.4   1.49GHz Transmission Line Oscillator

A microstrip transmission line oscillator, fabricated on RT Duroid $(\varepsilon_r = 10)$, is shown in Figure 4.12. The dimensions of the PCB are 50mm square.

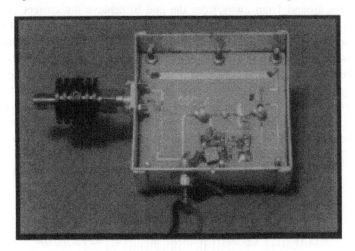

Figure 4.12   Transmission line oscillator

The transistor is a bipolar NE68135 $(I_C = 30\text{mA}, V_{CE} = 7.5\text{V})$. A 3dB Wilkinson power splitter is used to couple power to the external load. As mentioned earlier in Section 4.4, the output coupler causes a slight increase in amplifier noise figure. Phase compensation is achieved using a short length of transmission line and is finely tuned using a trimmer capacitor.

The oscillation frequency is 1.49GHz and αl is found to be 0.019 which sets $Q_0 = 83$. In these theories the absolute value of sideband noise power is independent of total output power so the noise power is quoted here both as absolute power and as the ratio with respect to carrier. Note that this is also a method for checking that saturation in the loop amplifier does not cause any degradation in performance [24]. The output power is 3.1dBm and the measured sideband noise power at 10kHz offset was −100.9dBm/Hz ± 1dB producing −104dBc. This is within 2dB of the theoretical minimum where the sideband noise is predicted to be −102.6dBm/Hz for a noise figure of 3dB and $Q_0 = 83$.

4.7.5   900MHz and 1.6GHz Oscillators Using Helical Resonators

Two low noise oscillators operating at 900MHz and 1.6GHz have been built using directly coupled helical resonators in place of the conventional transmission line resonators. These are built using the same topology as shown in Figure 4.11.

The structure of the copper L band helical resonators [22] with unloaded Qs of 350 to 600 is shown in Figure 4.13. and Figure 4.14. The helix produces both the central line and the shunt inductors; where the shunt inductors are formed by placing taps around 1mm away from the end to achieve the correct Q_L/Q_0. The equations which describe this resonator are identical to those used for the 'Fabry Pérot' resonator described earlier except for the fact that X now becomes $-Z_0/2\pi fl$ where l is the inductance and L is the effective length of the transmission line. As the Q becomes larger the value of the shunt l becomes smaller eventually becoming rather difficult to realize. The characteristic impedance of the helix used here is around 340Ω. It is interesting to note that this impedance can be measured directly using time domain reflectometry as these lines show low dispersion with only a slight ripple due to the helical nature of the line.

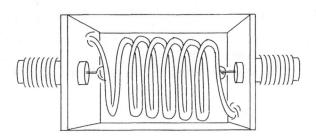

Figure 4.13 Helical resonator

Figure 4.14 Photograph of helical resonator

The SSB phase noise performance of the 900MHz oscillator was measured to be –127dBc/Hz at 25kHz offset for an oscillator with 0dBm output power, 6dB amplifier noise figure (Hybrid Philips OM345 amplifier), and Q_0= 582.

The 1.6GHz oscillator had a phase noise performance of -120dBc/Hz at 25kHz offset for Q_0 = 382, amplifier output power of 0dBm and amplifier noise figure of 3dB. The noise performance of both oscillators is within 2dB of the theoretical minimum noise performance available from an oscillator with the specified Q_0, Q_L and output power. In both cases the noise performance is 6dB lower at 50kHz offset demonstrating the correct $(1/\Delta f)^2$ performance.

4.7.6 Printed Resonators with Low Radiation Loss

Printed transmission line resonators have been developed consisting of a series transmission line with shunt inductors at either end as shown in Figure 4.15. Unloaded Q's exceeding 500 have been demonstrated at 4.5 GHz [23] and Qs of 80 at 22GHz on GaAs MMIC substrates. An interesting feature of these resonators is that they do not radiate and therefore do not need to be mounted in a screened box. This is due to the fact that the voltage nodes at the end of the resonator are minima greatly reducing the radiation losses.

Figure 4.15 Printed non-radiating high Q resonator

4.8   Tuning

In most oscillators there is a requirement to incorporate tuning usually by utilising a varactor diode. The theories described so far still apply but it is possible to gain a further insight by considering the power in the varactor as described by Underhill [25].

This is best illustrated by considering the two cases of narrow band tuning and broadband tuning.

4.8.1   Narrow Band Tuning

There are two main electronic methods to achieve narrow band tuning:

1.   Incorporate a varactor into the resonator.

2.   Incorporate a varactor based phase shift network into the feedback loop.

For narrow band tuning it is important to ensure the highest unloaded Q by adjusting the coupling of the varactor into the tuned resonant circuit. In fact for minimum effect the coupling should be as low as possible. The simplest method for achieving this is by utilising coupling capacitors in a parallel or series resonant network. These are of course dependent on the tuning range required and the varactor characteristics. The ratio of Q_L/Q_0 should be set as before to around 1/2 to 2/3 where Q_0 is now set by the varactor resonator loss combination   (For very narrow band applications it is often possible to ensure that the varactor does not degrade the resonator unloaded Q by ensuring very light coupling into the resonator).

4.8.2   Varactor Bias Noise

The only other requirement to consider is the noise on the varactor line which if large enough would degrade the inherent oscillator noise. Note that a flat noise spectral density on the varactor line will also produce a $1/\Delta f^2$ noise performance on the oscillator.

Note that a bias resistor $> 100\Omega$ is often large enough to generate sufficient thermal noise to degrade the phase noise performance of the oscillator and therefore only low value resistors should be used ($e_n^2 = 4kTBr_b$). Often in phase locked loop based oscillator systems an RC filter is used just prior to the oscillator to remove bias line noise. If this resistor is large (to make the C small) this will degrade the noise performance.

4.8.3   Tuning Using the Phase Shift Method

An alternative method for narrow band tuning is to incorporate an electronic phase shifter inside the feedback loop but separate from the resonator. The oscillator always operates at N.360° therefore narrow tuning can be incorporated using this phase shift. However, a degradation in noise occurs [24].

4.8.4   Degradation of Phase Noise with Open Loop Phase Error

If the open loop phase error of an oscillator is not close to Nx360°, the effective Q is reduced as the Q is proportional to the phase slope ($d\phi/d\omega$) of the resonator. Further the insertion loss of the resonator increases causing the closed loop amplifier gain to increase. It can be shown theoretically and experimentally that the noise performance degrades both in the thermal and flicker noise regions by $\cos^4\theta$ where θ is the open loop phase error. For high Q dielectric resonators, with for example a Q of 10,000, an offset frequency of 1MHz at 10GHz would produce a noise degradation of 6dB. A typical plot of the noise degradation with phase error for a high Q oscillator is shown in Figure 4.16 using both Silicon and GaAs active devices. This was measured by Cheng and Everard [24]. The full circles are for the GaAs devices and the open circles are for the silicon bipolar devices.

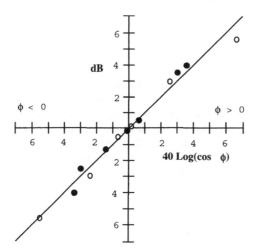

Noise performance degradation with open loop phase error. Bipolar - o,   GaAs -   • ,   Theory - ──────

Figure 4.16   Noise degradation with phase error

4.8.5   Broadband Tuning

The noise performance of a broad tuning range oscillator is usually limited by the
Q and the voltage handling capability of the varactor as has been described by
Underhill [25]. However this has not been applied to the oscillator operating under
optimum conditions. If it is assumed that the varactor diode limits the unloaded Q
of the total circuit, then it is possible to obtain useful information from a simple
power calculation. If the varactor is assumed to be a voltage controlled capacitor in
series with a loss resistor (R), The power dissipated in the varactor is:

$$P = \frac{V_R^{\,2}}{R} \tag{4.52}$$

The voltage across the capacitor V_C in a resonator is:

$$V_C = Q V_R \tag{4.53}$$

Therefore the power dissipated in the varactor is:

$$P_V = \frac{V_C^{\,2}}{Q^2 R} \tag{4.54}$$

The noise power in oscillators is proportional to $1/PQ_0^2$. Therefore the figure of
merit (V_C^2/R) should be as high as possible and thus the varactor should have large
voltage handling characteristics and small series resistance. However, the
definition of P and the ratio of loaded to unloaded Q are important and these will
alter the effect of the varactor on the noise performance. If we set the value of
Q_L/Q_0 to the optimum value where again the varactor defines the unloaded Q of the
resonator, then the noise performance of such an oscillator can be calculated
directly from the voltage handling and series resistance of the varactor. If the value
of Q_L/Q_0 is put in as 2/3 then:

$$L_{FM} = \frac{9FkTR_s}{16V_C^{\,2}} \left(\frac{f_0}{\Delta f} \right)^2 \tag{4.55}$$

If we take a varactor with a series resistance of 1Ω which can handle an RF voltage
of 0.25 volts rms at a frequency of 1GHz, then the noise performance at 25kHz
offset can be no better than -97dBc/Hz for an amplifier noise figure of 3dB. This
can only be improved by reducing the tuning range by more lightly coupling the
varactor into the tuned circuit, or by switching in tuning capacitors using PIN
diodes, or by improving the varactor. The voltage handling capability can be

improved by using two back to back diodes although care needs to be taken to avoid bias line currents.

4.8.6   Tunable 3.5 to 6GHz Resonator

A tunable resonator [26] [27] optimized to provide optimum conditions for low phase noise, over octave bandwidths, is shown in Figures 4.17 and 4.18. This consists of a transmission line with two variable reactances at either end. These consist of a series varactor and shunt inductor which vary the resonant frequency. As the transmission line has a low impedance in the middle of the line, at the operating frequency, the bias resistor can be made low impedance (50Ω). This means that the low frequency noise, which would cause unwanted modulation noise, can be kept low ($e_n^2 = 4kTBr_b$). The resistor also suppresses the second (and even) harmonic responses of the resonator.

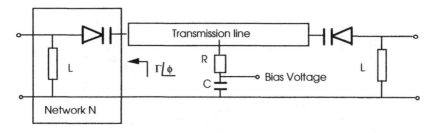

Figure 4.17 Tunable octave tuning transmission line resonator with near constant Q_L/Q_0
(3 to 6 GHz)

A prototype resonator operating from 3.5 to 6GHz was built on 25 thou alumina (ε_r=9.8) as shown in Figure 4.18. The varactor diodes are Alpha CVE7900D GaAs devices with C_{j0} = 1.5pf, Q (-4V, 50MHz) = 7000 breakdown voltage 45volts and k (capacitance ratio) = 6. The bias line resistor is 50Ω.

The measured response showing the variation of S_{21} and Q over the frequency range 3.5 to 6GHz is shown in Figure 4.19 and shows a loaded Q variation between 15 and 22 and insertion loss variation (S_{21}) between 3.5 to 6dB.

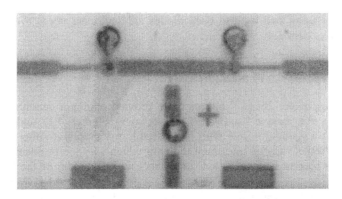

Figure 4.18 3.5 to 6GHz Resonator

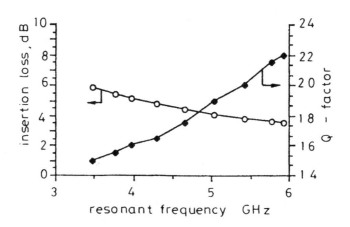

Figure 4.19 Variation of S_{21} and Q vs frequency

4.8.7   X Band Tunable MMIC Resonator

An MMIC version mounted on a 2 by 1mm GaAs substrate [27] is shown in Figure 4.20 and the response shown in Figure 4.21.

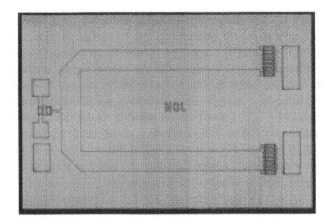

Figure 4.20 MMIC X band resonator

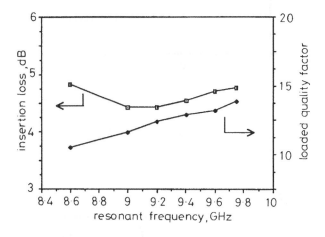

Figure 4.21 Variation of S_{21} and Q vs frequency

4.9   Flicker Noise Transposition

The theory described and the resulting optima apply only when the thermal (additive) noise is the major noise source. Here the noise in the oscillator falls off at a $1/\Delta f^2$ rate.

The effect of flicker noise can be incorporated into this theory by shaping part of the input noise with a $1/\Delta f$ characteristic on either side of the centre frequency as shown in Figure 4.22. Flicker noise occurs near DC where it is found that below a certain frequency (flicker corner, F_c) the noise power increases above the thermal noise at a rate approximately proportional to $1/f$. In an oscillator, this noise is transposed up to the operating frequency by non-linear operating conditions. The oscillator loop transfer function remodifies the output signal, producing a $(1/\Delta f)^3$ characteristic over the band in which flicker noise occurs. It should be noted, however, that there is a transposition gain/loss which means that ΔF_c is not necessarily equal to F_c. This transposition gain/loss is dependent on the nonlinear operating conditions of the oscillator including power level, amplifier impedances, biasing conditions and waveform shape [3][15][29][34][35][41]. Hajimiri and Lee [42] have shown that ΔF_c is always less than F_c and that waveform symmetry is important.

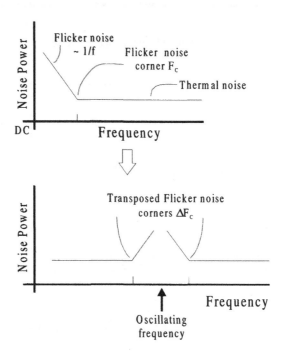

Figure 4.22 Flicker noise transposition

Present understanding for GaAs based devices suggests that PM noise is produced by noise modulation from the gate series noise voltage onto the input non-linear capacitance and that AM noise is caused by modulation of the channel width. In an oscillator however these mechanisms become intertwined due to AM to PM conversion.

In fact, for flicker noise (modulation noise) as described by Parker [8] from within the active device and within the resonator as described by Montress, Parker and Loboda [9], the same optima do not apply and it is often the case that Q_L should be as high as possible. This is because thermal noise is additive noise and therefore the sideband noise does not vary with total power level. However modulation noise sidebands increase with total power. It is therefore important to investigate methods to reduce the effect of flicker noise.

4.10   Current Methods for Transposed Flicker Noise Reduction

A number of methods have been devised to reduce this problem without changing the active device. These include:

1.   RF detection and LF cancellation.

2.   Direct LF reduction.

3.   Transposed gain amplifiers and oscillators.

4.   Transposed flicker noise suppression using feedforward amplifiers.

4.10.1   RF Detection and LF Cancellation

Z. Galani, M.J. Bianchini, R.C. Waterman, R. Dibiase, R.W Laton and J.B. Cole [10] (1984), detected the phase noise by using a two port resonator as both the oscillator resonator and as a one port PM discriminator as shown in Figure 4.23. The signal from the discriminator was then used to apply cancellation to a separate phase modulator within the loop. A 20dB suppression of close to carrier noise was reported at 10kHz offset from a carrier at 10GHz.

Ivanov, Tobar and Woode [32] [33] have produced some of the lowest noise room temperature X band oscillators using sapphire dielectric resonators incorporating error correction. The resonators operate in a whispering gallery mode with an unloaded Q of 180,000. A further enhancement has been incorporated into the error correcting circuit, which now incorporates an interferometer and a low noise GaAs microwave amplifier as shown in Figure 4.24. The interferometer has the effect of carrier suppression, thereby reducing the power through the amplifier which reduces the effect of the flicker noise degradation caused by this amplifier. This is because the flicker noise is modulation noise which would produce lower power sidebands for lower carrier levels. The microwave amplifier also reduces the effect of the noise figure of the mixer.

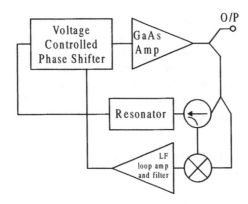

Figure 4.23   RF detection and LF cancellation, Z. Galani , M.J. Bianchini, R.C. Waterman,
R. Dibiase, R.W Laton and J.B. Cole [10]

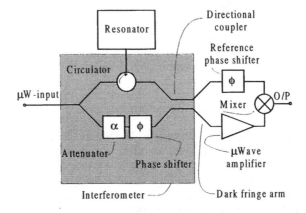

Figure 4.24 Interferometer for phase noise reduction, Ivanov, Tobar and Woode [32] [33]

This form of carrier suppression is an extension of the technique used by
Santiago and Dick [36] who used carrier suppression to improve the stability of a
microwave oscillator using a cryogenic sapphire discriminator [37].

Driscoll and Weinert [11] developed a technique which detected the noise at
the input and the output of the amplifier and applied LF cancellation as shown in
Figure 4.25. Both these techniques have been shown to be successful at removing
flicker noise in oscillators.

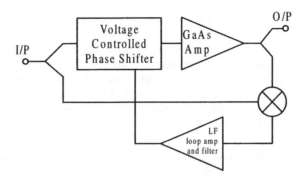

Figure 4.25 RF detection and LF cancellation, Driscoll and Weinert [11]

4.10.2 Direct LF Reduction

Noise reduction was discussed by Riddle and Trew [12], (1985), who designed the feedback amplifier using a pair of FETs operated in push pull at the microwave frequency, but operated in parallel at low frequencies via an LF connection between the two bias networks.

Pringent and Obregon [13], (1987), used a bias network with LF negative feedback. This reduced the device gain at low frequencies and at the same time reduced the baseband and transposed flicker noise. This assumed that the majority of the flicker noise was generated by a gate noise source modulating the input non-linear capacitor of the GaAs FET.

An elegant implementation of the same idea was produced by Mizzukami et al [14] (1988), who developed a GaAs MMIC in which the impedance presented to the source was arranged to rise at LF. This method would be more difficult to implement with discrete FETs, as the parasitics need to be very low indeed.

These direct LF reduction methods have reduced the LF flicker noise present at the device terminals, but this often does not necessarily correlate well with the oscillator flicker noise reduction. The transposed flicker noise depends on the nature of the internal noise sources, and the transposition mechanism. All of these vary greatly between device manufacturers.

Walls, Ferre-Pikal and Jefferts [34] and Ferre-Pikal, Walls and Nelson [35] have investigated, analytically, the transfer function of bipolar transistors to investigate whether the flicker modulation processes can be reduced. Using this information they produced amplifiers with residual flicker corners around 5Hz. As yet these amplifiers have not been used in oscillators.

4.10.2.1   Cross Correlation Flicker Noise Measurement System

To enable a better understanding of this process Dallas and Everard [15], [16] developed a measurement system, initially presented at the 1990 IEEE MTT Conference, capable of measuring the cross correlation coefficients between the baseband noise on the drain and the AM and PM components transposed onto the carrier. Martinez, Oates and Compton [28] presented a similar system for measurements on a P-HEMT. These systems are extensions of the phase bridge arrangements presented by Riddle and Trew [30] and Sann [31].

Further measurements have now been made which include noise measurements on the gate as well and describe techniques for deconvolving the internal noise sources [29]. These measurements indicate why low frequency feedback often does not greatly improve the flicker noise, as the correlation between the baseband noises at the terminals and the demodulated noise on the carrier is often low.

The system is shown in Figure 4.26. A low noise reference signal is passed through the amplifier and both the gate and drain noise are measured. The AM and PM components of the noise are measured using the delay line discriminator either in phase for AM or at 90 degrees for PM. By measuring the direct noise at the gate and drain of the device and the demodulated noise simultaneously on a digitizing card the cross correlation functions can be derived. From these measurements Dallas and Everard has shown that the internal noise sources and their correlation coefficients can be derived [29]. From these measurements new techniques for Flicker noise reduction have been devised. The final system had a residual noise floor of -175 dBc for offsets greater than 3 kHz.

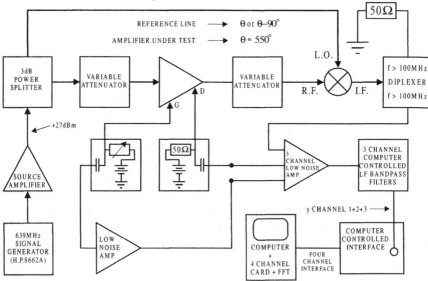

Figure 4.26   Flicker noise measurement System

A typical cross correlation function vs delay is shown in Figure 4.27 which, in fact, demonstrated a higher correlation for AM rather than PM for this particular device.

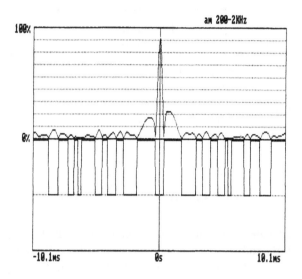

Figure 4.27   AM/Drain Cross Correlation Function

4.10.3   Transposed Gain Oscillators

Everard and Page-Jones [17] [18] [19] and Driscoll and Weinert [11] developed independently the transposed gain amplifier and oscillator as an alternative means for reducing flicker noise. The system proposed by Everard and Page-Jones also included a delay line in between the two LO paths to ensure negligible phase noise degradation from the LO. This enables the use of an LO with poor phase noise. In the oscillators described later the suppression between L_{FMTGA} and L_{FMLO} was measured to be greater than 60dB. This ensured that the LO phase noise did not degrade the total oscillator phase noise at all. Care does need to be taken,however, to ensure that drift in the LO does not vary the mixer performance.

A Transposed Gain Oscillator is shown in Figure 4.28. The transposed gain amplifier consists of two silicon Schottky barrier diode mixers and a RF amplifier operating up to 200 MHz. A GaAs based LO drives both mixers to transpose the gain of the 200 MHz amplifier to ± 200 MHz of LO frequency. By using silicon mixers and a silicon based RF amplifier the flicker noise corner is set by these components to be around 1kHz. The noise on the GaAs LO is rejected by inserting a delay line which matches the delay of the RF amplifier. With a differential delay of less than 10 ps this gives > 60dB suppression at 10 kHz offset.

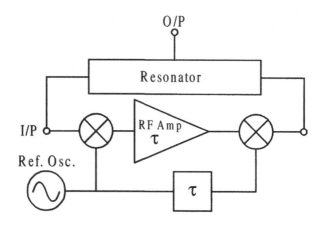

Figure 4.28 Transposed gain oscillator

A transposed gain oscillator is shown in Figure 4.29 [17] [18] [19]. This operates at 7.6GHz and uses two Watkins Johnson silicon Schottky barrier diode mixers (MZ 7420C) driven with 20dBm of LO power. Semi-rigid cable and variable phase shifters are used to provide the correct delays in all the arms. The resonator is temperature stabilised using heater elements to ±1 mK as the temperature coefficient for these resonators within the cavity was 40ppm/K.

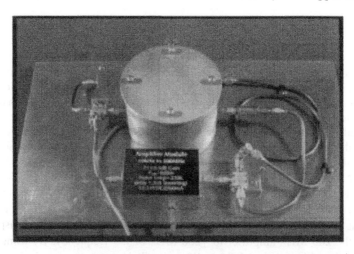

Figure 4.29 Transposed gain oscillator

The oscillators use room temperature sapphire resonators operating in the fundamental $TE_{01\delta}$ mode. These resonators are operated substantially as a two port resonator with a lightly coupled third port to provide a filtered output. The unloaded Q (including the output port) is measured to be 44,000 by using the relationship $S_{21} = (1 - Q_L/Q_0)$ and measuring the insertion loss and Q. The resonator cavity is shown in Figure 4.30 and consists of a sapphire ½ inch puck mounted on a castellated quartz tube (to reduce contact area and maintain high Q. A UV curing adhesive was used to bond the quartz support to the sapphire with low loss.

Figure 4.30 Sapphire resonator operating at 7.6GHz

The theory used to design these oscillators, used the equations which incorporated the power available at the output definition. Therefore under optimum operating conditions:

$$L_{FM} = \frac{FkT}{8Q_0^2 \left(Q_L/Q_0\right)^2 \left(1 - Q_L/Q_0\right)^2 P_{AVO}} \left(\frac{f_0}{\Delta f}\right)^2 \qquad (4.32)$$

when $Q_L/Q_0 = 1/2$:

$$L_{FM} = \frac{2FkT}{Q_0^2 P_{AVO}} \left(\frac{f_0}{\Delta f}\right)^2 \qquad (4.38)$$

To obtain minimum noise performance, the insertion loss of the resonator is set to be 6dB producing a loaded Q of 22,000. The insertion loss of the mixers is around 7 dB; therefore the open loop gain of the silicon amplifier should exceed 20 dB.

The total noise figure of the two mixers and amplifier was around 15 dB including the effect of image noise.

To measure the noise performance two 7.5 GHz oscillators were built using high level mixers and beat together to produce an IF around 15MHz. This was then locked to an HP8662A signal generator which incorporated a divide by 8 to reduce the phase noise of the signal generator.

The wanted sideband output power obtained from the O/P mixer was around +8.0dBm. The new noise measurements show a phase noise performance of::

1.   -136dBc/Hz @ 10 kHz offset (theory -139).

2.   -114dBc/Hz @ 1kHz offset.

Phase noise measurements of these oscillators are within 3dB of the theoretical minimum at 10kHz offset and within 5dB of the theoretical minimum at 1kHz, if flicker noise is not included in the theory. Similar oscillators using $BaTiO_3$ resonators with unloaded Qs of 24,000 have been produced by Broomfield and Everard with a noise performance of -130dBc/Hz at 10kHz offset. This is about 6 dB worse than using the sapphire resonators, as would be expected from the $1/Q_0^2$ in the theory.

There are two limitations with this system.

1.   The maximum output power is limited by the mixer to around 8dBm.

2.   The raised noise figure is around 15dB.

4.10.4   Transposed Flicker Noise Suppression Using Feedforward Amplifiers in Oscillators

Recently Broomfield and Everard presented reduced flicker noise transposition in a 1watt feedforward amplifier at the IEEE 2000 Frequency Control Symposium [45]. In [45] it was shown that the feedforward technique could be used to reduce the transposed flicker noise (in an amplifier) in a manner similar to the way it reduces distortion and conventional thermal noise. This was demonstrated in [45] using phase and amplitude characterisation of the amplifier.

A simplified feedforward amplifier is shown in Figure 4.31. G is the amplifier gain, K_1 is the coupler ratio. The main path without error correction goes from A to B. The error signal is derived from the left hand loop AC–AD. The error correction is then obtained from the upper path and the right hand loop AB–EB.

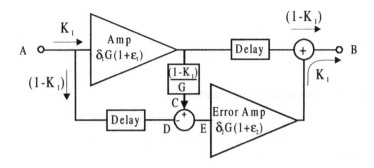

Figure 4.31 Feedforward amplifier topology

Flicker noise is modulation noise which means that the ratio of sideband noise to carrier level is constant. Therefore the absolute level of the sideband noise varies with carrier level.

A phasor analysis is performed in [45], however a simplified analysis is performed here. The flicker noise of the main amplifier is represented as $(1 + \varepsilon_1)$ and the error correcting amplifier as $(1 + \varepsilon_2)$. The amplifier balance is represented as δ_1 and δ_2. Note that perfect balance occurs when $\delta = 1$ therefore the suppression of each loop in decibels $= 10\log(1-\delta)$. The power transfer function of the amplifier shown in Figure 4.31 is therefore:

$$P_T = K_1 G(1 - K_1)[\delta_1 \varepsilon_1(1 - \delta_2) + \delta_2 \varepsilon_2(1 - \delta_1) - \delta_1 \delta_2 \varepsilon_1 \varepsilon_2 + (\delta_1 - \delta_1 \delta_2 + \delta_2)] \quad (4.56)$$

where $K_1 G(1 - K_1)$ is the ideal amplifier gain. Assuming that δ_1 and δ_2 are close to one (of order 0.1% to 1%) and ε is very small, then this equation simplifies to:

$$P_T = K_1 G(1 - K_1)[1 + \varepsilon_1(1 - \delta_2) + \varepsilon_2(1 - \delta_1)] \quad (4.57)$$

If each loop suppression is assumed to be 23dB and the flicker noise levels of each amplifier the same then a flicker noise suppression of 20dB can be obtained. Of course the total noise cannot be suppressed below the noise figure of the feedforward amplifier.

4.10.4.1 Noise Figure Analysis

In feedforward amplifiers, all noise components originating from the main amplifier are suppressed and the noise figure of the system is set by the noise produced in the error correcting amplifier and the losses in that signal path [46] as shown in equation (3). The noise figure for our topology is 11.5dB.

$$F_{ff} = \frac{F_{error}}{L_{input\ coupler} L_{delay\ line} L_{error\ amp\ input\ coupler}} \tag{4.58}$$

4.10.4.2 Amplifier gain requirements

From oscillator phase noise analysis it has been shown that if the system phase noise is thermal noise limited, the overall single sideband phase noise of the oscillator can be expressed as:

$$L_{FM} = \frac{FkT}{8Q_0{}^2 (Q_L/Q_0)^2 (1 - Q_L/Q_0)^2 P_{AVO}} \left(\frac{f_0}{\Delta f}\right)^2 \tag{4.32}$$

where F = the amplifier noise figure, k = Boltzmann constant, T = temperature, Q_0 = the unloaded Q of the resonator, Q_L = the loaded Q of the resonator, P_{AVO} = the power available from the output of the amplifier, f_0 = the centre frequency and Δf = the carrier offset frequency. The noise performance is therefore a minimum when $Q_L/Q_0 = \frac{1}{2}$ and the amplifier gain is 4 (6 dB). Excluding coupler losses and assuming $K_1 = \frac{1}{2}$ the ideal amplifier gain should therefore be 16 (12dB). To allow for other losses 15dB was chosen.

4.10.4.3 Amplifier Realisation

A 1watt feedforward power amplifier that operates over a ±250MHz bandwidth at 7.6GHz has been designed and built [45]. This system utilised two commercially available multi-stage GaAs based amplifiers, both with gains of 15dB and a P_{1dB} of +35dBm for the main amplifier and a P_{1dB} of +28dBm for the error-correcting amplifier. The directional couplers, hybrid combiners, power dividers, attenuators and delay lines have all been designed and produced in house. The signal loops are balanced by allowing the amplifiers to reach a quasi-stable temperature state and adjusting both phase and amplitude within the loops, for maximum cancellation.

4.10.4.4 Residual Flicker Noise Measurement

The residual flicker noise of the amplifier was measured using a delay line frequency discriminator and a reduction in residual phase noise level of approximately 20 dB was observed (Figure 4.32).

The flicker noise level below the carrier was calculated to be less than – 150dBc/Hz at 10kHz offset when the error correcting amplifier was switched on.

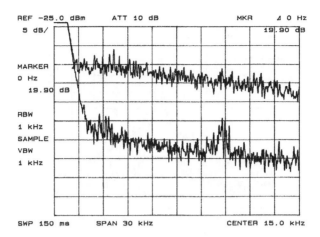

Figure 4.32 Residual flicker noise measurement: upper trace, error correcting amplifier switched off; lower trace, error correcting amplifier switched on. This shows ~20dB suppression over the measured bandwidth of 1 to 30kHz.

4.10.4.5 The Feedforward oscillator

The feedforward amplifier can be used as the oscillation sustaining stage in a low noise oscillator [47] [48]. However, the amplifier must not be allowed to saturate as this will affect the loop balance and effectiveness of the amplifier to suppress flicker noise. An oscillator is therefore produced with the addition of a resonator, an output coupler, a phase shifter and a limiter, as shown in Figure 4.33.

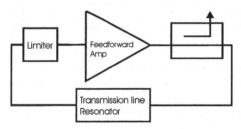

Figure 4.33 Diagram of the feedforward oscillator

A 7.6GHz transmission line resonator was designed with 2.5dB insertion loss and a loaded Q of 45 ($Q_0 \sim 180$), a 10dB output coupler and a silicon diode limiter with a threshold level of +9dBm. A typical phase noise plot can be seen in figure 4.34 and a phase noise comparison between the oscillator with the error correction on, and error correction off, is shown in table 4.1. The reduced value of Q_L/Q_0 was required to account for the losses of the limiter and loop components.

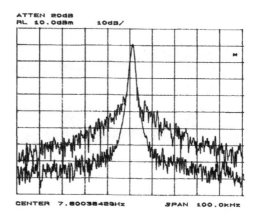

Figure 4.34 Phase noise measurement: upper trace, error correction off; lower trace, error correction on.

Table 4.1 A comparison of single sideband phase noise when the error correction is switched on and off

Single side band phase noise level, L (dBc/Hz)			
Offset from carrier	Error amp OFF	Error amp ON	Theory
12.5kHz	-79.8	-100.3	-102.5
25kHz	-88.5	-107.5	-108.5
50kHz	-98.6	-114.4	-114.5
100kHz	-107.2	-120.8	-120.5

Assuming the noise figure of the feedforward amplifier is 11.5 dB (1.3dB for limiter, 5dB loss for couplers and delay line and 5.2dB for error amplifier – all measured), P_{AVO} = +16dBm and Q_L/Q_0 = 1/4 then L_{FM} = –107.5dBc/Hz at 25kHz offset which is within 1 dB of the theory. It should be noted that the close to carrier phase noise approximately follows a $1/f^3$ law when the error correction is switched off, and a $1/f^2$ law when the error correction is switched on.

4.11   Non-linear CAD

A very brief introduction to non-linear analysis will be described. It is often difficult, however, to know how to probe an oscillator as it is an autonomous self-sustaining system. Therefore a simple technique which enables a break at any point within the oscillator circuit is presented here. This was used with Volterra series analysis to predict output frequency, output, power at the fundamental and harmonics and biasing conditions [40].

The technique requires four steps which are illustrated in Figure 4.35.

1.   Break the circuit at a short circuit point.

2.   Insert a current source and frequency dependent resistor at this point.

3.   Make resistor: Open circuit at the fundamental.
                            Short circuit at the harmonics.

4.   Adjust the amplitude and fundamental frequency of the current source to obtain zero volts.

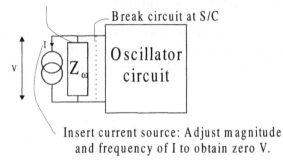

Figure 4.35   Method for non-linear CAD of oscillating systems

Step 4 is performed iteratively to obtain a voltage lower than a specified value, at which point the current obtained is a solution to the circuit. In the process of calculating this current all the other important parameters of the circuit would have been calculated.

4.12   Summary for Minimum Phase Noise

The theory required to design low phase noise oscillators using most of the commonly used resonators has been described and a number of experimental circuits presented. The major parameters which set the noise performance have been described. These are:

1.   High unloaded Q and low noise figure.

2.   The coupling of the resonator set to achieve: $Q_L/Q_0 = 1/2$ to $2/3$.

3.   Use of a device and circuit configuration producing the lowest transposed flicker noise corner ΔF_C.

4.   The open loop phase error set to be N.360°.

5.   For tuning, incorporation of the varactor loss resistor into the loss of the resonator and Q_L/Q_0 set as before.

6.   For narrow band tuning, loose coupling of the varactor into the resonator and Q_L/Q_0 set as before, or consideration of the use of a low loss phase shifter in the feedback loop. Expect 6dB noise degradation if the open loop phase error goes to 45°.

7.   Arranging for low bias line noise.

4.13   Detailed Design Example

Design a 150 MHz oscillator using a 235nH inductor with a Q_0 of 300. Use an inverting amplifier with an input and output impedance of 50Ω. An LC resonator with losses can therefore be represented as an *LCR* resonator as shown in Figure 4.36.

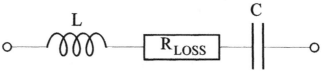

Figure 4.36 Model of *LC* resonator including losses

As:

$$Q_0 = \frac{\omega L}{R_{LOSS}}$$

the equivalent series resistance is 0.74Ω. Including the output impedance, the ratio Q_L/Q_0 is:

$$\frac{Q_L}{Q_0} = \frac{R_{LOSS}}{R_{LOSS} + R_{IN} + R_{OUT}}$$

For $R_{IN} = R_{OUT}$:

$$\frac{Q_L}{Q_0} = \frac{R_{LOSS}}{R_{LOSS} + 2R_{IN}}$$

Let:

$$\frac{Q_L}{Q_0} = \frac{1}{2}$$

then $R_{LOSS} = 2\,R_{IN}$

$$R_{IN} = \frac{R_{loss}}{2}$$

Therefore $R_{IN} = R_{OUT} = 0.37\Omega$ as shown in Figure 4.37.

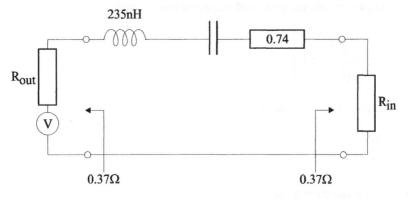

Figure 4.37 *LC* resonator with scaled source and load impedances

As the amplifier has a 50Ω input and output impedance, impedance transformers are required. Use *LC* transformers as shown in Figure 4.38.

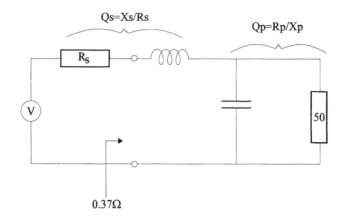

Figure 4.38 *LC* transformer to convert to 50Ω

The equations for the series and shunt components are:

$$Q_S = Q_p = \sqrt{\left(\frac{R_p}{R_S} - 1\right)}$$

The Q of the series component is:

$$Q_S = \frac{X_s}{R_s}$$

The Q of the shunt component is:

$$Q_p = \frac{R_p}{X_p}$$

Note:

1. R_p = shunt resistance

2. R_s = series resistance

3. X_s = series reactance = $j\omega L$

4. X_p = shunt reactance = $1/j\omega C$

$$Q_S = Q_p = \sqrt{\left(\frac{R_p}{R_s}-1\right)} = \sqrt{\left(\frac{50}{0.37}-1\right)} = 11.58$$

$X_s = 4.28 = j\omega L$ therefore $L = 4.5$ nH

$X_p = 4.31 = 1/j\omega C$, therefore $C = 246$ pf

Incorporate two transforming circuits into the resonator circuit as shown in Figure 4.39.

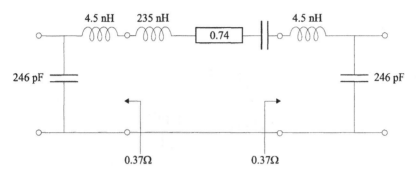

Figure 4.39 LC resonator with impedance transformers

As the total inductance is 235nH, the inductors within this matching network can be incorporated into this main inductor. The part which resonates with the series capacitor is therefore reduced by 9nH as shown in Figure 4.40.

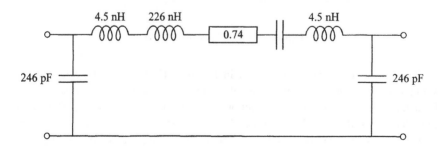

Figure 4.40 Resonator with total $L = 235$nH

It is now necessary to calculate the resonant frequency. The part of the inductance which resonates with the series capacitance is reduced by the matching inductors to: 235nH - (2 x 4.5 nH)   = 226 nH

As:

$$LC = \frac{1}{(2\pi f)^2}$$

$$C = \frac{1}{L(2\pi f)^2} = 5pf$$

The circuit is now as shown in Figure 4.41:

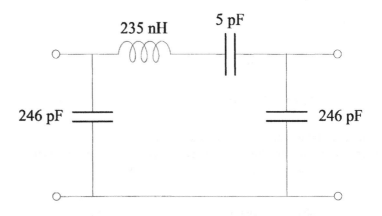

Figure 4.41 Final resonator circuit

The frequency response is shown in Figure 4.42. It should be noted that the shunt capacitors are very large and therefore the effect of any series parasitic inductance in these inductors is important. The effect of adding 1nH and 2nH inductances is illustrated in Figure 4.43. The effect of incorporating a shunt inductor of two nH can be removed by reducing the shunt capacitors from 246pf to 174pf.

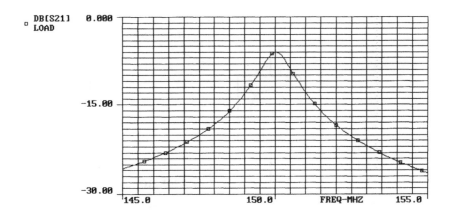

Figure 4.42 Frequency response of resonator

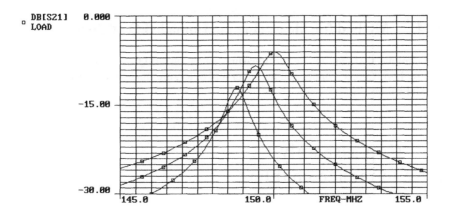

Figure 4.43 The effect of parasitic inductance: Upper trace, no parasitic; middle trace, 1nH parasitic inductor; lower trace, 2nH parasitic inductor

Note for 6dB insertion loss, the phase shift at resonance is 180°. So the amplifier should provide a further 180°. If necessary a phase shifter should be included to ensure N x 360° at the peak in the resonance as shown in Figure 4.44.

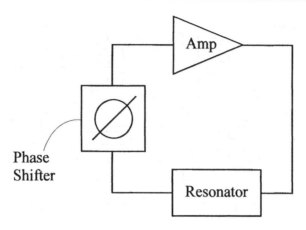

Figure 4.44 Oscillator incorporating phase shifter

4.14   Method for Measuring the Unloaded Q of Coils

A technique that can be used to measure the unloaded Q of coils is illustrated in
Figure 4.45. Loops, connecting the inner and outer, are made at the end of a pair of
coaxial cables.

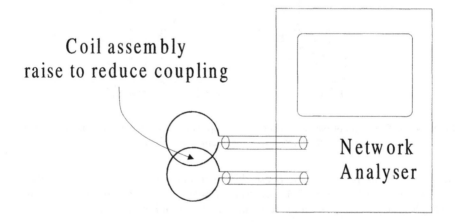

Figure 4.45 Method Used to Measure Unloaded Q of Coils

The two loops are overlaid and the overlap adjusted to obtain a minimum in the
coupling coefficient using a network analyser The coil to be measured is then

parallel resonated with a high Q capacitor and the combination used to lightly couple power from one loop to the other. The Q is calculated using the equation:

$$Q_o = \frac{f_o}{2\Delta f}$$

where $2\Delta f$ is the frequency difference between the 3dB points and f_0 is the centre frequency. The measured value of Q will only be correct if the coupling coefficient is kept low, thereby avoiding loading due to the loops and the network analyser.

4.15   References

1. T.E. Parker, "Current Developments in SAW Oscillator Stability", *Proceedings of the 31st Annual Symposium on Frequency Control*, Atlantic City, New Jersey, 1977, pp. 359–364.

2. J.K.A. Everard, "Low Noise Power Efficient Oscillators: Theory and design", *Proceedings of the IEE, Pt G*, **133**, No. 4, pp. 172–180, 1986.

3. J.K.A. Everard, "Low Noise Oscillators", Chapter 8 in D.G. Haigh and J.K.A. Everard (eds), *GaAs Technology and its Impact on Circuits and Systems, IEE*, Peter Peregrinus, 1989.

4. J.K.A. Everard, "Minimum Sideband Noise in Oscillators", *Proceedings of the 40th Annual Symposium on Frequency Control*, Philadelphia, Pennsylvania, 28–30 May 1986, pp. 336–339.

5. W.P. Robins, *Phase Noise in Signal Sources*, IEE, Peter Peregrinus, 1982.

6. J.K.A. Everard and J. Bitterling, "Low Phase Noise Highly Power Efficient Oscillators", *IEEE International Frequency Control Symposium*, Orlando, Florida, 27–30 May 1997.

7. P.A. Moore and S.K. Salmon, "Surface Acoustic Wave Reference Oscillators for UHF and Microwave Generators", *Proceedings of the IEE*, Pt H, **130**, No. 7, pp. 477–482.

8. T.E. Parker, "Characteristics and Sources of Phase Noise in Stable Oscillators", *Proceedings of the 41st Annual Frequency Control Symposium*, 1977, pp. 99–100.

9. G.Montress, T.E. Parker and M.J. Loboda, "Residual Phase Noise Measurements of VHF, UHF, and Microwave Components", *Proceedings of the 43rd Annual Frequency Control Symposium*, 1989.

10. Z. Galani, M.J. Bianchini, R.C. Waterman, R. Dibiase, R.W. Laton and J.B. Cole, "Analysis and Design of a Single-resonator GaAs FET Oscillator with Noise Degeneration", *IEEE Transactions on Microwave Theory and Techniques*, **MTT-32**, No. 12, pp. 1556–1565, 1984.

11. M.M. Driscoll and R.W. Weinert, "Spectral Performance of Sapphire Dielectric Resonator-controlled Oscillators Operating in the 80K to 275K Temperature Range", *IEEE Frequency Control Symposium*, San Francisco, June 1995, pp. 410–412.

12. A.N. Riddle and R.J. Trew, "A New Method of Reducing Phase Noise in GaAs FET Oscillators", *IEEE MTT-S Digest*, pp. 274–276, 1984.

13. M. Pringent and J. Obregon, "Phase Noise Reduction in FET Oscillators by Low-frequency Loading and Feedback Circuitry Optimization", *IEEE Transactions on Microwave Theory and Techniques*, **MTT-35**, No. 3, pp. 349–352, 1987.

14. H. Mizzukami et al., "A High Quality GaAs IC Tuner for T.V./V.C.R. Receivers", *IEEE Transactions on Consumer Electronics*, Vol. CE-34, No. 3, pp. 649–659, 1988.

15. P.A. Dallas and J.K.A. Everard, "Measurement of the Cross Correlation between Baseband and Transposed Flicker Noises in a GaAs MESFET", *IEEE Microwave Theory and Techniques Conference*, Dallas, Texas, May 1990, pp. 1261–1264.

16. P.A. Dallas, *Determining the Sources of Flicker Noise in GaAs MESFETs*, PhD Thesis, King's College, London, 1995.

17. J.K.A. Everard and M.A. Page-Jones, "Ultra Low Noise Microwave Oscillators with Low Residual Flicker Noise", *IEEE International Microwave Symposium*, Orlando, Florida, 16–20 May 1995, pp. 693–696.

18. J.K.A. Everard and M.A. Page-Jones, "Transposed Gain Microwave Oscillators with Low Residual Flicker Noise", *IEEE Frequency Control Symposium*, San Francisco, June 1995, pp. 374–378.

19. M.A. Page-Jones and J.K.A. Everard, "Enhanced Transposed Gain Microwave Oscillators", *European Frequency and Time Forum*, Brighton, 5–7 March 1996, pp. 275–278.

20. G.Montress, T.E. Parker, M.J. Loboda and J.A. Greer, "Extremely Low Phase Noise SAW Resonators and Oscillators: Design and performance", *IEEE Transactions on Ultrasonics, Ferroelectrics and Frequency Control*, **35**, No. 6, pp. 657–667, 1988.

21. J.K.A. Everard and K.K.M. Cheng, "Novel Low Noise 'Fabry Perot' Transmission Line Oscillator", *IEE Electronics Letters*, **25**, No. 17, pp. 1106–1108, 1989.

22. J.K.A. Everard, K.K.M. Cheng and P.A. Dallas, "A High Q Helical Resonator for Oscillators and Filters in Mobile Communications Systems", *IEE Electronics Letters*, **25**, No. 24, pp. 1648–1650, 1989.

23. J.K.A. Everard and K.K.M. Cheng, "High Performance Direct Coupled Bandpass Filters on Coplanar Waveguide", *IEEE Transactions on Microwave Theory and Techniques*, **MTT-41**, No. 9, pp. 1568–1573, 1993.

24. K.K.M. Cheng and J.K.A. Everard, "Noise Performance Degradation in Feedback Oscillators with Non-zero Phase Error", *Microwave and Optical Technology Letters*, **4**, No. 2, pp. 64–66, 1991.

25. M.J. Underhill, "Oscillator Noise Limitations", *IERE Conference Proceedings 39*, 1979, pp. 109–118.

26. K.K.M. Cheng and J.K.A. Everard, "Novel Varactor Tuned Transmission Line Resonator", *IEE Electronics Letters*, **25**, No. 17, pp. 1164–1165, 1989.

27. K.K.M. Cheng and J.K.A. Everard, "X Band Monolithic Tunable Resonator/Filter", *IEE Electronics Letters*, **25**, No. 23, pp. 1587–1589, 1989.

28. R.D. Martinez, D.E. Oates and R.C. Compton, "Measurements and Model for Correlating Phase and Baseband 1/f Noise in an FET", *IEEE Transactions on Microwave Theory and Techniques*, **MTT-42**, No. 11, pp. 2051–2055, 1994.

29. P.A. Dallas and J.K.A. Everard, "Characterisation of Flicker Noise in GaAs MESFETs for Oscillator Applications", *IEEE Transactions on Microwave Theory and Techniques*, **MTT-48**, No. 2, pp. 245–257, 2000.

30. A.N. Riddle and R.J. Trew, "A New Measurement System for Oscillator Noise Characterisation", *IEEE MTT-S Digest*, 1987, pp. 509–512.

31. K.H. Sann, "The Measurement of Near Carrier Noise in Microwave Amplifiers", *IEEE Transactions on Microwave Theory and Techniques*, **16**, September, pp. 761–766, 1968.

32. E.N. Ivanov, M.E. Tobar and R.A. Woode, "Ultra-low-noise Microwave Oscillator with Advanced Phase Noise Suppression System", *IEEE Microwave and Guided Wave Letters*, **6**, No. 9, pp. 312–314, 1996.

33. E.N. Ivanov, M.E. Tobar and R.A. Woode, "Applications of Interferometric Signal Processing to Phase-noise Reduction in Microwave Oscillators", *IEEE Transactions on Microwave Theory and Techniques*, **MTT-46**, No. 10, pp. 1537–1545, 1998.

34. F.L. Walls, E.S. Ferre-Pikal and Jefferts, "Origin of 1/f PM and AM Noise in Bipolar Junction Transistor Amplifiers", *IEEE Transactions on Ultrasonics, Ferroelectrics, and Frequency Control*, **44**, March, pp. 326–334, 1997.

35. E.S. Ferre-Pikal, F.L. Walls and C.W. Nelson, "Guidelines for Designing BJT Amplifiers with Low 1/f AM and PM Noise", *IEEE Transactions on Ultrasonics, Ferroelectrics, and Frequency Control*, **44**, March, pp. 335–343, 1997.

36. D.G. Santiago and G.J. Dick, "Microwave Frequency Discriminator with a Cooled Sapphire Resonator for Ultra-low Phase Noise", *IEEE Frequency Control Symposium*, May 1992, pp. 176–182.

37. D.G. Santiago and G.J. Dick, "Closed Loop Tests of the NASA Sapphire Phase Stabiliser", *IEEE Frequency Control Symposium*, May 1993, pp. 774–778.

38. M.C.D. Aramburo, E.S. Ferre-Pikal, F.L. Walls and H.D. Ascarrunz, "Comparison of 1/f PM Noise in Commercial Amplifiers", *IEEE Frequency Control Symposium, Orlando, Florida*, 28–30 May 1997, pp. 470–477.

39. J.K.A. Everard, "A Review of Low Phase Noise Oscillator Design", *IEEE Frequency Control Symposium, Orlando, Florida*, 28–30 May 1997, pp. 909–918.

40. K.K.M. Cheng and J.K.A. Everard, "A New and Efficient Approach to the Analysis and Design of GaAs MESFET Microwave Oscillators", *IEEE International Microwave Symposium, Dallas, Texas*, 8–16 May 1990, pp. 1283–1286.

41. F.M. Curley, *The Application of SAW Resonators to the Generation of Low Phase Noise Oscillators*, MSc Thesis, King's College, London, 1987.

42. A. Hajimiri and T.H. Lee, "A General Theory of Phase Noise in Electrical Oscillators", *IEEE Journal of Solid State Circuits*, **33**, No. 2, pp. 179–194, 1998.

43. D.B. Leeson, "A Simple Model of Feedback Oscillator Noise Spectrum", *Proceedings of the IEEE*, **54**, February, pp. 329–330, 1966.

44. L.S. Cutler and C.L. Searle, "Some Aspects of the Theory and Measurement of Frequency Fluctuations in Frequency Standards", *Proceedings of the IEEE*, **54**, February, pp. 136–154, 1966.

45. C.D. Broomfield and J.K.A. Everard, "Flicker Noise Reduction Using GaAs Microwave Feedforward Amplifiers", *IEEE International Frequency Control Symposium*, Kansas City, June 2000.

46. N. Pothecary, *Feedforward Linear Power Amplifiers*, Artech House, 1999, pp. 113–114.

47. J.K.A. Everard and C.D. Broomfield, *Low Noise Oscillator*, Patent application number 0022308.1.

48. J.K.A. Everard and C.D. Broomfield, "Transposed Flicker Noise Suppression in Microwave Oscillators Using Feedforward Amplifiers", *IEE Electronics Letters*, **26**, No. 20, pp. 1710–1711.

5

Mixers

5.1   Introduction

Mixers are used to translate a signal spectrum from one frequency to another. Most modern RF/microwave transmitters, receivers and instruments require many of these devices for this frequency translation. The typical symbol for a mixer is shown in Figure 5.1.

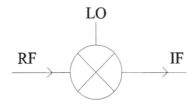

Figure 5.1 Typical symbol for a mixer

An ideal mixer should multiply the RF and LO signals to produce the IF signal. It should therefore translate the input spectrum from one frequency to another with no distortion and no degradation in noise performance. Most of these requirements can be met by the perfect multiplication of two signals as illustrated in equation (5.1):

$$(V_1 \cos \omega_1 t)(V_2 \cos \omega_2 t) = \frac{V_1 V_2}{2}(\cos(\omega_1 + \omega_2)t + \cos(\omega_1 - \omega_2)t) \quad (5.1)$$

Here it can be seen that the output product of two input frequencies consists of the sum and difference frequencies. The unwanted sideband is usually fairly easy to remove by filtering. Note that no other frequency terms other than these two are generated. In real mixers there are a number of compromises to be made and these will be discussed later.

Mixing is often achieved by applying the two signals to a non linear device as shown in Figure 5.2.

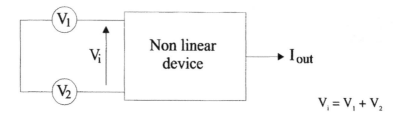

$$V_i = V_1 + V_2$$

Figure 5.2 Mixing using a non-linear device

The non linearity can be expressed as a Taylor series:

$$I_{out} = I_0 + a[V_i(t)] + b[V_i(t)]^2 + c[V_i(t)]^3 + ...$$ (5.2)

Taking the squared term:

$$b(V_1 + V_2)^2 = b(V_1^2 + 2V_1V_2 + V_2^2)$$ (5.3)

It can immediately be seen that the square law term includes a product term and therefore this can be used for mixing. This is illustrated in Figure 5.3 where the square law term and the exponential term of the diode characteristics are shown.

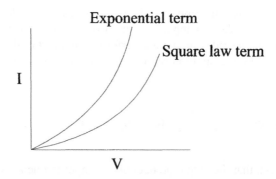

Figure 5.3 Diode characteristic showing exponential and square law terms

Note of course that there are other terms in the equation which will produce unwanted frequency products, many of them being in band. Further, as the signal voltages are increased the difference between the two curves increases showing that there will be increasing power in these other unwanted terms.to achieve this non-linear function a diode can be used as shown in Figure 5.4:

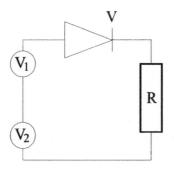

Figure 5.4 Diode operating as a non-linear device

If two small signals are applied then multiplication will occur with rather high conversion loss. The load resistor could also include filtering. If the LO is large enough to forward-bias the diode then it will act as a switch. This is a single ended mixer which produces the wanted signal and both LO and RF breakthrough. It can be extended to a single balanced switching action as shown in figure 5.5.

5.2   Single balanced mixer (SBM)

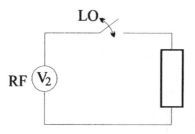

Figure 5.5 Switching single balanced mixer

The waveform and therefore operation of this switching mixer are now shown to illustrate this slightly different form of operation which is the mode of operation used in many single balanced transistor and diode mixers. The LO switching waveform has a response as shown in Figure 5.6.

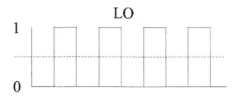

Figure 5.6 LO waveform for SBM

The spectrum of this is shown in equation (5.4) and consists of a DC term and the odd harmonics whose amplitude decreases proportional to 1/n.

$$S(t) = \frac{1}{2} + \sum_{n=1}^{\infty} \frac{\sin(n\pi/2)}{n\pi/2} \cos(n\omega_0 t)$$ (5.4)

If this LO signal switches the RF signal shown in figure 5.7 then the waveform shown in Figure 5.8 is produced.

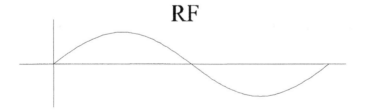

Figure 5.7 RF signal for DBM

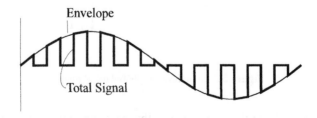

Figure 5.8 Output waveform for SBM

The spectrum of this can be seen to produce the multiplication of the LO (including the odd harmonics, with the RF signal. This produces the sum and difference frequencies required as well as the sum and difference frequencies with each of the odd harmonics. The output voltage is therefore:

$$V_0(t) = V_{RF}(t) \times S(t)$$

$$= V_{RF} \cos \omega_{RF} t. \left[\frac{1}{2} + \sum_{n=1}^{\infty} \frac{\sin(n\pi/2)}{(n\pi/2)} \cos(n\omega_{LO} t) \right] \qquad (5.5)$$

It is important to note that there is no LO component. There is however an RF term due to the product of V_{RF} with the DC component of the switching term. This shows the properties of an SBM in that the LO term is rejected. It is often useful to suppress both the LO and RF signal and therefore the DBM was developed.

5.3   Double Balanced Mixer (DBM)

If the switch is now fed with the RF signal for half the cycle and an inverted RF signal for the other half then a double balanced mixer (DBM) is produced. This is most easily illustrated in Figure 5.9.

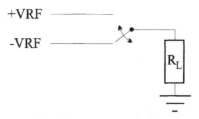

Figure 5.9 Switching double balanced mixer

The output voltage is given by:

$$V_0(t) = 2V_{RF} \cos \omega_{RF} t \left[\sum_{n=1}^{\infty} \frac{\sin(n\pi/2)}{(n\pi/2)} \cos(n\omega_{LO} t) \right] \qquad (5.6)$$

The waveform is shown in Figure 5.10.

Envelope

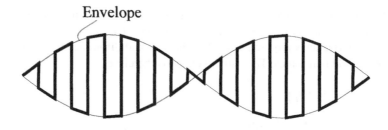

Figure 5.10 Output waveform for switching DBM

Note that there is now no LO or RF breakthrough although the odd harmonics still appear, but these are usually filtered out as shown in Figure 5.11.

Envelope

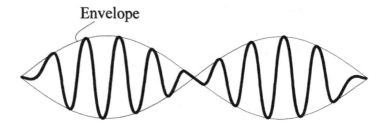

Figure 5.11 Filtered wave form from DBM

Note also that in practice there will be some breakthrough, typically around 50dB at LF degrading to 20 to 30 dB at VHF and UHF.

5.4   Double Balanced Transistor Mixer

A double balanced transistor mixer is shown in Figure 5.12. This is the standard 'Gilbert Cell' configuration. The RF input is applied to the base of transistors Q_1 and Q_2. For correct operation these devices should not be driven into saturation and therefore signal levels considerably less than the 1 dB compression point should be used. This is around 12mV rms if there are no emitter degeneration resistors. For third order intermodulation distortion better than 50 dB the RF drive level should be less than around 2 mV rms which is around –40 dBm into 50Ω. (as in the NE/SA603A mixer operating at 45 MHz). This requirement for a low level input is a very important characteristic of most transistor mixers.

The LO is applied to the base of Q_3, Q_4, Q_5 and Q_6. and these transistors provide the switching action. Gains of 10 to 20 dB are typical with noise figures of 5 dB at VHF going up to 10 dB at 1GHz.

The collectors of Q_1 and Q_2 provide the positive and negative V_{RF} as previously shown in Figure 5.9. Q_3 and Q_5 switch between them to provide the RF signal or the inverted RF signal to the left hand load. Q_4 and Q_6 switch between them for the right hand load. The output should be taken balanced between the output loads.

An LF version of the Gilbert Cell double balanced mixer is the 1496 which is slightly different in that the lower transistors Q_1 and Q_2 are each fed from separate current sources and a resistor is connected between the emitters to set the gain and emitter degeneration. This greatly increases the maximum RF signal handling capability to levels as high as a few volts.

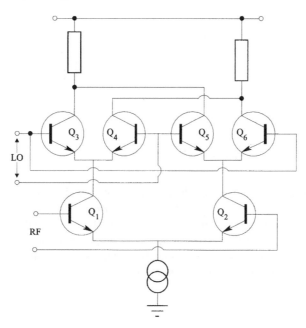

Figure 5.12 Double balanced transistor mixer

5.5   Double Balanced Diode Mixer

The double balanced diode mixer is shown in Figure 5.13. The operation of this mixer is best described by looking at the mixer when the LO is either positive or negative as shown in Figures 5.14 and 5.15. When the LO forward biases a pair of diodes, both can be represented as resistors. Note that the RF current flows through both resistors and good balance requires that these resistors should be the same.

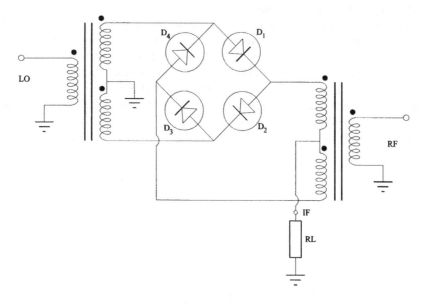

Figure 5.13 Double balanced diode mixer

When the LO is positive (the dots are positive) and diodes D_1 and D_2 become conducting and connect the 'non-dot' arm of the RF transformer to the IF port. In the next half cycle of the LO, diodes D_3 and D_4 will conduct and connect the 'dot arm' of the RF transformer to the IF port. The output therefore switches between the RF signal and the inversion of the RF signal, which is the requirement for double balanced mixing as shown earlier in Figure 5.9.

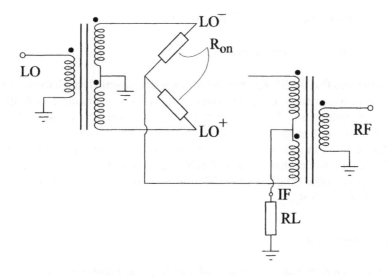

Figure 5.14 LO causes D_3 and D_4 to conduct

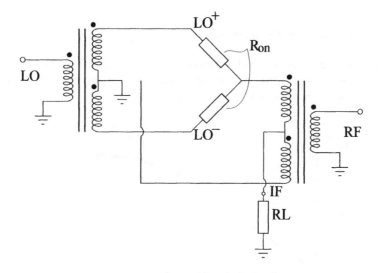

Figure 5.15 LO causes D_1 and D_2 to conduct

5.6   Important Mixer Parameters

5.6.1   Single Sideband Conversion Loss or Gain

This is the loss that the RF signal experiences when passing through the mixer. For a double balanced diode mixer the single sideband loss is around 5 to 8 dB. The theoretical minimum is 3 dB as half the power is automatically lost in the other sideband. The rest of the power is lost in the resistive losses in the diodes and transformers and in reflections due to mismatch at the ports. The noise figure is usually slightly higher than the loss. Double balanced transistor mixers often offer gain of up to around 20 dB at LF/VHF with noise figures of 5 to 10 dB at 50MHz and 1GHz respectively.

5.6.2   Isolation

This is the isolation between the LO, RF and IF ports. Feedthrough of the LO and RF components is typically around -50dB at LF reducing to -20 to -30dB at GHz frequencies.

5.6.3   Conversion Compression

This defines the point at which conversion deviates from linearity by a certain amount. For example, the 1dB compression point is the point at which the conversion loss increases by 1dB (Figure 5.16).

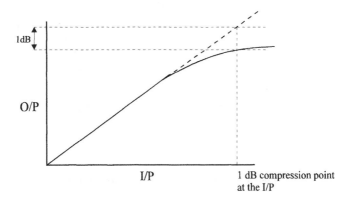

Figure 5.16 Gain compression

5.6.4   Dynamic Range

This is defined as the amplitude range over which the mixer provides correct performance. Dynamic range is measured in dB and is the input RF power range over which the mixer is useful. The lower limit of the dynamic range is set by the noise power and the higher level is set by the 1dB compression point or the intermodulation performance specification required.

5.6.5   Two Tone Third Order Intermodulation Distortion

If the input to the RF port consists of two tones then it is found that third order intermodulation distortion is a critical parameter. This distortion is caused by the cubic term in the expansion of the diode non-linearity, see equation (5.2), as shown in:

$$C\left(V_{RF1} + V_{RF2} + V_{LO}\right)^3 \tag{5.7}$$

and produces unwanted output terms within the desired band. If two tones are applied to the RF port, they should produce 2 IF output tones at f_1 and f_2. Third order intermodulation distortion produces signals at $2f_1 - f_2$ and $2f_2 - f_1$. For example if two signals at 100MHz + 10kHz and 100MHz + 11kHz are incident on the RF port of the mixer and the LO is 100MHz, the output will consist of two tones at 10kHz and 11kHz. Third order intermodulation will produce two unwanted tones at 9kHz and 11kHz. Further, because these signals are third order signals they increase in power at three times the rate of the wanted output signal. Therefore an increase in 1dB in the wanted signal power causes a 3dB increase in the unwanted signal power degrading the distortion to wanted signal level by 2 dB.

This form of distortion is therefore very important and needs to be characterised when designing mixer systems. The concept of third order intercept point was therefore developed.

5.6.6   Third Order Intercept Point

The intercept point is a theoretical point (extrapolated) at which the fundamental and third order response intercept. This is illustrated in Figure 5.17.

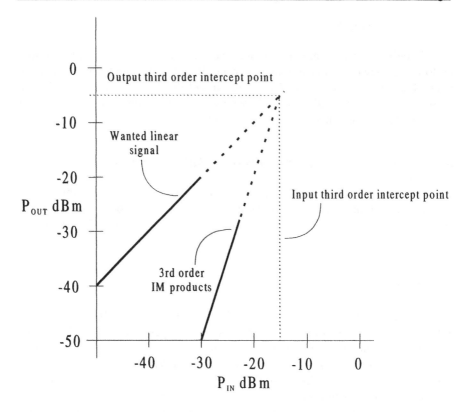

Figure 5.17 Third order intercept

This point is a concept point where the mixer could not actually operate, but it offers a technique which can be used to obtain the value of distortion signal levels at lower power levels. The intercept point can be defined either at the input or at the output but here we will refer to the **input** intercept point. The intermodulation distortion level is therefore:

$$P_{im} = [P_{ITC} - 3(P_{ITC} - P_{RF})] \tag{5.8}$$

The difference between the input RF level and the distortion level is therefore:

$$P_{RF} - P_{im} = 2 (P_{ITC} - P_{RF}) \tag{5.9}$$

Take an example. If the RF drive level is at 0dBm and intercept point at +20dBm. The third order line goes down by 3 x 20 = 60 dB. Therefore the difference is 40dB.

5.6.7   LO Drive Level

This is the LO drive level required to provide the correct operating conditions and conversion loss. It varies typically from +7dBm to +22dBm for double balanced mixers. Mixers designed to operate at high power levels with lower distortion often use more than one diode in each arm therefore requiring higher LO power to switch. Lower drive levels can be achieved by using a DC bias.

5.7   Questions

1.   A mixer with an LO drive level of +7dBm has a third order (input) intercept point of +15dBm. Calculate the signal power required to achieve a third order distortion ratio better than 20dB, 40dB and 60dB.

2.   Design a 150 ±50 MHz to 800 ±50MHz converter using a double balanced mixer. The system is required to have a signal to noise ratio and signal to third order intercept ratio greater than 40 dB. What are the maximum and minimum signal levels that can be applied to the mixer? The mixer is assumed to have a loss and noise figure of 6dB and a third order (input) intercept point of +10dBm. Note that thermal noise power in a 1 Hz bandwidth is $kT = -74$dBm/Hz.

3.   A spectrum analyser is required to have a third order spurious free range of 90 dB. What is the maximum input signal to guarantee this for a mixer with +10dBm third order (input) intercept point?

Note therefore that when testing distortion on a spectrum analyser that it is important to check the signal level at the mixer.

5.8   Bibliography

1. W.H. Hayward, Introduction to Radio Frequency Design Prentice Hall 1982.

2. H.L. Krauss, C.W. Bostian and F.H. Raab, Solid state Radio Engineering Wiley 1980.

6

Power Amplifiers

6.1   Introduction

Power amplifiers consist of an active device, biasing networks and input and output reactive filtering and transforming networks. These networks are effectively bandpass filters offering the required impedance transformation. They are also designed to offer some frequency shaping to compensate for the roll-off in the active device frequency response if broadband operation is required. However the function of each network is quite different. The input circuit usually provides impedance matching to achieve low input return loss and good power transfer. However, the output network is defined as the 'Load Network' and effectively provides a load to the device which is chosen to obtain the required operating conditions such as gain, efficiency and stability. For example if high efficiency is required the output network should not match the device to the load as match causes at least 50% of the power to be lost in the active device. Note that maximum power transfer using matching causes 50% of the power to be lost in the source impedance.

For high efficiency one usually requires two major factors: optimum waveform shape and a device output impedance which is significantly lower than the input impedance to the load network.

This chapter will provide a brief introduction to power amplifier design and will cover:

1.   Load pull measurement and design techniques.

2.   A design example of a broadband efficient amplifier operating from 130 to 180 MHz.

3.   A method for performing real time large signal modelling on circuits.

When power amplifiers are designed the small signal S parameters become modified owing to changes in the input, feedback and output capacitances due to changes in g_m due to saturation and charge storage effects. It is therefore often useful to obtain large signal parameters through device measurement and modelling. It is also important to decide on the most important parameters in the design such as:

1.   Power output.

2.   Gain.

3.   Linearity.

4.   The VSWR and load phase over which the device is stable.

5.   DC supply voltage.

6.   Efficiency.

To this end many amplifiers are designed using load pull large signal techniques. This technique offers:

1.   Measurement of the device under the actual operating conditions including the correct RF power levels.

2.   Correct impedance matching at the input and orrect load network optimisation for the relevant operating conditions incorporating the effect of large signal feedback.

3.   Both CW and pulsed measurements and designs.

4.   Measurement for optimum power out and efficiency.

These load pull techniques also incorporate jigs which enable accurate large signal models to be developed. These models can then be used in a more analytical way to design an amplifier as illustrated in the design example.

6.2   Load Pull Techniques

A system for making load pull measurements is shown in Figure 6.1 and offers both CW and pulsed measurements. The system consists of a signal generator and a directional coupler on the input side of the device jig. The directional coupler measures both the incident and reflected power with a typical coupling coefficient of -around -20dB dB. Note the signal generator often incorporates an isolator to prevent source instability and power variation as the load is varied.

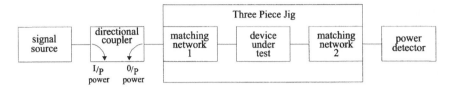

Figure 6.1 Large signal measurement set-up

The signals are then applied to a three stage jig capable of being split into three parts as shown in Figure 6.2. This jig includes an input matching network, a device holder and an output matching network and can be split after the measurement. The matching networks could include microstrip matching networks, *LC* matching networks and transmission line tuneable stub matching networks. The printed matching networks are varied using silver paint.

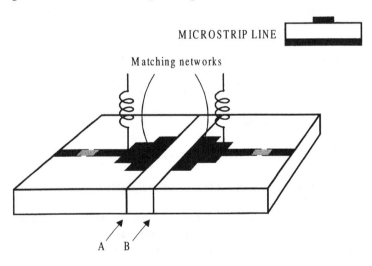

Figure 6.2 Three piece large signal jig

The output of the amplifier jig is connected to a power detector. In fact the power detector for either the input or output of the directional coupler could consist of a modern spectrum analyser. Most modern analysers are capable of both CW and pulsed power measurements.

The procedure is as follows:

1. Bias the device and allow to stabilise. Monitor temperature and ensure adequate heat sinking. Use forced air cooling or water cooling if necessary.

2. Monitor the current provided as this is important to prevent device damage and for calculation of efficiency.

3. Apply an input signal to the amplifier and adjust the input and output matching networks iteratively using either silver paint or trimmer capacitors or sliding stub tuners. These should be adjusted to obtain both low reflected power from the input and the required output power and efficiency. Note that this is an iterative process and it is quite possible to obtain non optimum maxima. The variable matching networks should also be arranged to offer a reasonable tuning range of impedance.

4. Now split the jig and while terminating the input and output in 50Ω measure the impedance into the jigs from the device end (A or B). For good input match this reverse impedance is the complex conjugate of the amplifier input impedance. Note however that the impedance looking into the output jig (from the device) is not necessarily a match but it is the correct load impedance to obtain the required output power, gain and efficiency.

5. Now redesign the input and output matching networks to obtain all the required components centred on reasonable values and then test the amplifier again. Also remove any external stubs that were used by incorporating their operation into the amplifier matching networks. Note that the losses in the initial matching network may have been significant so the second iteration may produce slightly different results.

6. Test for stability by varying the bias and supply voltage over the full operating range and by applying a variable load network capable of varying the impedance seen by the amplifier.

Note that it is also possible to use a similar test jig with 50Ω lines to measure the small signal S parameters while varying the bias conditions and thereby deduce a large signal equivalent circuit model. This is used in the design example given to produce an efficient broadband power amplifier. Although the techniques are applied to a Class E amplifier they are equally applicable to any of the amplifier classes.

6.3   Design Example

The aim here is to design a power amplifier covering the frequency range 130 MHz to 180 MHz.

6.3.1   Introduction

This section describes techniques whereby broadband power-efficient Class E amplifiers, with a passband ripple of less than 1dB, can be designed and built. The amplifiers are capable of operating over 35% fractional bandwidths with efficiencies approaching 100%. As an example, a 130 to 180 MHz Class E amplifier has been designed and built using these techniques. At VHF frequencies the efficiency reduces owing to the non-ideal switching properties of RF power devices. A large signal model for an RF MOS device is therefore developed, based on DC and small signal S parameter measurements, to allow more detailed analysis of the Class E amplifier. The non-linearities incorporated in the model include the non-linear transconductance of the device, including the reverse biased diode inherent in the MOS structure, and non-linear feedback and output capacitors. The technique used to develop this model can be applied to other non-linear devices. Close correlation is shown between experimental and CAD techniques at 150 MHz. CAD techniques for rapidly matching the input impedance of the non-linear model are also presented.

6.3.2   Switching Amplifiers

High efficiency amplification is usually achieved by using a switching amplifier where the switch dissipates no power and all the power is dissipated in the load. An ideal switching device dissipates no power because it has no voltage across it when it is on, no current flowing through it when it is off, and zero transition times.

In real switching amplifiers there are three main loss mechanisms:

1.   The non-zero on-resistance of the switching device.

2.   The simultaneous presence of large voltages and currents during the switching transitions.

3.   The loss of the energy stored in the parasitic shunt capacitance of the switching device at switch-on.

The losses caused by the on-resistance can be reduced by ensuring that the on-resistance is considerably less than the load resistance presented to the switching device.

The switching transition losses can be reduced by choosing a device with a fast switching time. Efficiency can be further increased if the overlap of the voltage and current waveforms can be reduced to minimise the power losses during the switching transitions. The loss of the energy stored in the shunt capacitance at switch-on ($\frac{1}{2}CV^2$) can be reduced by choosing a device with a low parasitic shunt capacitance. At VHF even a small parasitic shunt capacitance can result in large losses of energy. The requirement to discharge this capacitor at switch-on also imposes secondary stress on the switching device.

6.3.3   ClassE Amplifiers

The Class E amplifier proposed by the Sokals [7] [8] [9] and further analysed by Raab [10] [11] is designed to avoid discharging the shunt capacitance of the switching device and to reduce power loss during the switching transitions. This is achieved by designing a load network for the amplifier, which determines the voltage across the switching device when it is off, to ensure minimum losses. Typical Class E amplifier waveforms are shown in Figure 6.3.

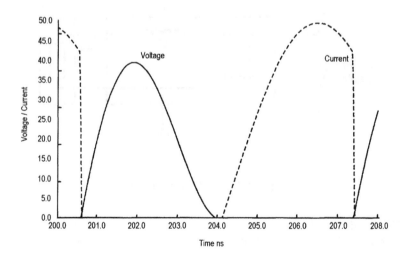

Figure 6.3 Class E amplifier voltage and current waveforms

The design criteria for the voltage are that it:

1.   Rises slowly at switch-off.

2.   Falls to zero by the end of the half cycle.

3.   Has a zero rate of change at the end of the half cycle.

A slow rise in voltage at switch-off reduces the power lost during the switch off transition. Zero voltage across the switching device at the end of the half-cycle ensures that there is no charge stored in the parasitic shunt capacitance when it turns on, so that no current is discharged through the device. Zero rate of change at the end of the half cycle reduces power loss during a relatively slow switch-on transition by ensuring that the voltage across the switching device remains at zero while the device is switching on.

The circuit developed by the Sokals (Figure 6.4) uses a single switching device (BJT or FET) and a load network consisting of a series tuned LC network (L_2, C_2) an RF choke (L_1) and a shunt capacitor (C_1) which may be partly or wholly made up of the parasitic shunt capacitance of the switching device.

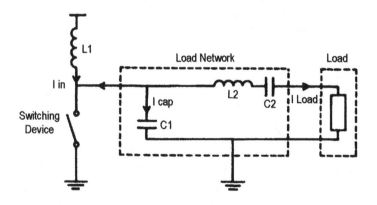

Figure 6.4 Basic Class E amplifier

The RF choke (L_1) is sufficiently large to provide a constant input from the power supply. The series LC circuit (L_2, C_2) is tuned to a frequency lower than the operating frequency and can be considered, at the operating frequency, as a series tuned circuit in series with an extra inductive reactance. The tuned circuit ensures a substantially sinusoidal load current (Figure 6.5) and the inductive reactance causes a phase shift between this current and the fundamental component of the applied voltage. The difference between the constant input current and the sinusoidal load current flows through the switching device when it is on and through the shunt capacitor (C_1) when it is off. The capacitor/switch current is therefore also sinusoidal; however, it is now 180° out of phase with respect to the load current and contains a DC offset to allow for the current flowing through the RF choke (L_1).

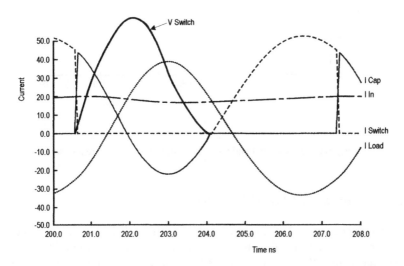

Figure 6.5 Class E current waveforms F_c = 147 MHz

As the voltage across the switching device when it is off is the integral of the current through the shunt capacitor (C_1), the phase shift introduced by the LC circuit adjusts the point at which the current is diverted from the switch to the capacitor. This ensures that the voltage waveform (Figure 6.5) meets the criteria for Class E operation by integrating the correct portion of the offset sinusoidal capacitor current. This point is determined by ensuring that the integral of the capacitor current over the half-cycle is zero and that the capacitor current has dropped to zero by the end of the half-cycle.

Owing to the fact that the *LC* series tuned circuit is tuned to a frequency which is lower than the operating frequency, the conventional Class E amplifier has a highly frequency dependent amplitude characteristic (Figure 6.6). It is this change of impedance which prevents optimum Class E operation from being achieved over a wide bandwidth.

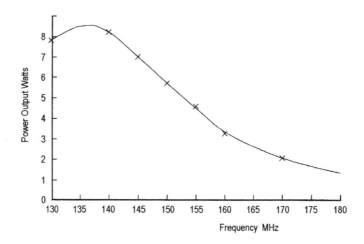

Figure 6.6 Narrow band Class E amplifier frequency response

6.3.4 Broadband Class E Amplifiers

To enable the design of broadband Class E amplifiers [14] a closer examination of the voltages and currents of the narrow band amplifier is required (Figure 6.5). As the voltage across the switch is defined by the integral of the current flowing through the shunt capacitor (C_1), and as the AC component of this current also flows through the series LC circuit (L_2, C_2) when the switch is off, then the load angle of the series tuned circuit defines the optimum angle for producing the correct voltage waveform. This load angle defines the phase shift between the fundamental components of the voltage across the switch and the current flowing through the series tuned circuit (L_2, C_2). In the basic Class E amplifier circuit the harmonic impedance of the series tuned circuit is assumed to be high because of its Q. The value of the shunt capacitor (C_1) must also be correct to produce the correct voltage when the switch is off and to satisfy the steady state conditions. The load angle of the total network is also therefore important.

Simulation and analysis of an ideal narrow band circuit (Figure 6.6) shows that the load angle of the off-tuned series tuned circuit should be 50°and the angle of the total circuit should be 33°.

If the load network is designed without incorporating a shunt capacitor a simple broadband network can be designed. This should be designed with a greater load angle (50°), which reduces to the required 33° when a shunt capacitor is added. The slope in susceptance with frequency caused by this capacitor is removed as described later.

A circuit capable of presenting a constant load angle over a very large bandwidth is shown in Figure 6.7 and its susceptance diagram in Figure 6.8.

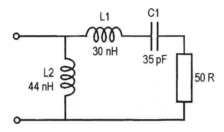

Figure 6.7 Broadband load angle network

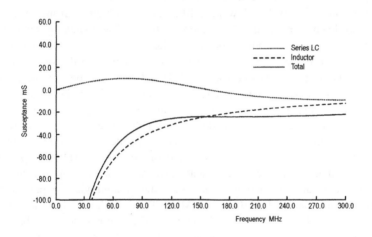

Figure 6.8 Susceptance of broadband circuit

The circuit consists of a low Q series tuned circuit in parallel with an inductor. At the resonant frequency of the tuned circuit the slope of its susceptance curve is designed to cancel the slope of the susceptance curve of the inductor. This allows the circuit to maintain a constant susceptance over a wide bandwidth. An analysis of this circuit is given in Section 6.3.9 at the end of this chapter and it is shown that optimum flatness can be achieved when:

$$L_2 = R/\omega \tan 50° \qquad (6.1)$$

$$C_1 = 2L_2/R^2 \qquad (6.2)$$

$$L_1 = 1/\omega^2 C_1 \qquad (6.3)$$

The load network impedance at the harmonics should be purely reactive to ensure no losses in the load network. The impedance should also be fairly high to ensure that the integral of the current through the shunt capacitor, and hence the current through the capacitor, should be similar to that in the narrow band circuit to meet the criteria for Class E operation. It should also be high to avoid harmonic power being dissipated in the on-resistance of the switching device.

To reduce the power output at the harmonics this circuit was combined with a broadband matching network and a third order bandpass filter to produce a circuit which presented a load angle of 50° over the band 130MHz to 180 MHz. The filter was based on a Chebyshev low pass filter design, obtained from Zverev [12], which had been converted to a bandpass filter. The matching network was arranged to increase the output power of the amplifier by decreasing the load presented to the device. The final network was designed to deliver 12 W into a 50Ω load using a 12V power supply. This network can be considered as the broadband equivalent of the off-tuned series *LC* circuit of the simple Class E amplifier, which presents a load angle of 50° at the operating frequency.

When the shunt capacitor is added to this network a slope in susceptance is introduced owing to the capacitor's frequency dependence. A slope was therefore placed on the impedance curve of the network using a CAD AC optimisation package so that when the shunt capacitance was included in the network it presented a constant 33° load angle and a constant magnitude of input impedance of 12Ω over the band 130 MHz to 180 MHz. The impedance of the load network without the capacitor now has a load angle of 50° in the middle of the band which slightly increases at higher frequencies and slightly reduces at lower frequencies.

During AC optimisation it was found that a number of components could be removed without degrading the performance. The final network and its impedance are shown in Figures 6.9 and 6.10.

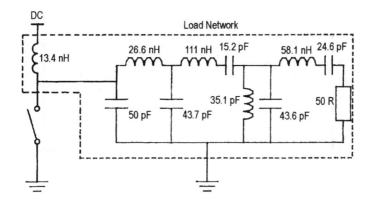

Figure 6.9 Broadband load network

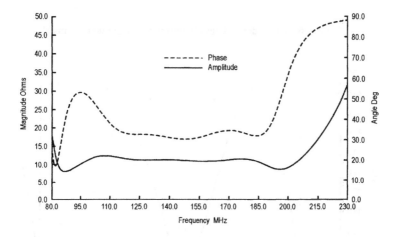

Figure 6.10 Broadband load network impedance

The broadband amplifier was then simulated in the time domain (Figures 6.11 and 6.12) using a switch with the following characteristics:

1. 1Ω on-resistance.

2. $1\ M\Omega$ off-resistance.

3. 1 ns switching time.

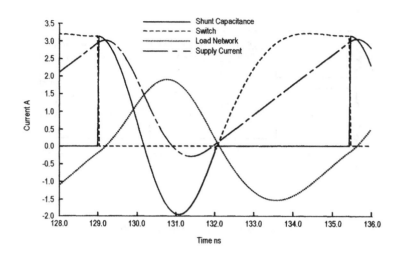

Figure 6.11 Broadband Class E amplifier current waveforms

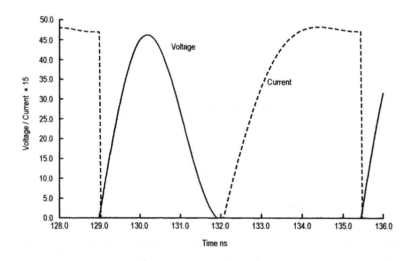

Figure 6.12 Broadband Class E amplifier voltage and current waveforms

The simulation showed that the amplifier, with a 12 V supply, was capable of delivering 12W (11dBW) into a 50Ω load with 85% efficiency over the band 130MHz to 180 MHz (Figures 6.13 and 6.14). A discrete fourier transform showed that second, third and fourth order harmonics were all better than 45 dB below the fundamental.

The current waveforms (Figures 6.11 and 6.12) are different from those for the simple Class E amplifier (Figures 6.3 and 6.5)) with the exception of the current through the shunt capacitor while the switching device is off. As this is the same, the voltage across the device (Figure 6.12) is the same as for the simple Class E amplifier and the criteria for maximum efficiency described earlier are met. The RF choke is now part of the load network and therefore the input current is now an asymmetric sawtooth.

6.3.5   Measurements

A broadband amplifier was constructed and the impedance of the load network was checked with a network analyser. A power MOSFET was used as the switching device; this FET has a parasitic shunt capacitance of approximately 35pF, so an additional 22pF trimmer was placed in parallel with the device to achieve the required capacitance. The measured results (Figures 6.13 and 6.14) show that the output power remained fairly constant over the band at approximately half the value for the simulated amplifier. The efficiency remained fairly constant at approximately 60% and a power gain of 10dB was achieved with a 0.5W drive power. The input matching network of the constructed amplifier was not broadband and was adjusted for a perfect match using a directional coupler at each frequency measurement. A broadband input matching network could be designed to achieve a complete broadband amplifier.

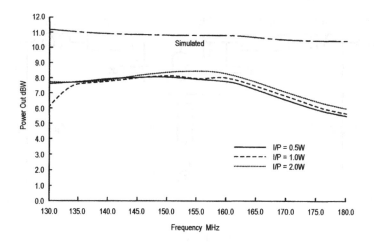

Figure 6.13 Class E broadband amplifier. Graph of power out vs frequency

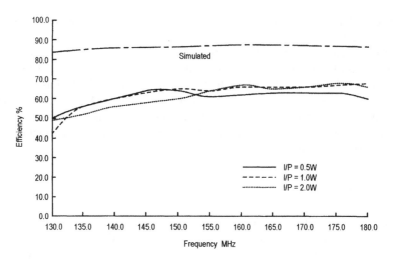

Figure 6.14 Class E broadband amplifier measurements. Graph of efficiency vs frequency

6.3.6   Non-linear Modelling

To enable more accurate analysis of the experimental amplifier a non-linear model of the active MOS device was developed. The circuit used for the large signal model is shown in Figure 6.15 where the component values are determined by measurements from an actual device.

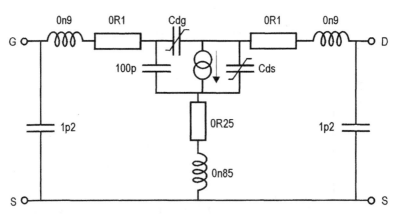

Figure 6.15 MOSFET model

A test jig was built for the power MOSFET and the DC characteristics of the device were measured on a curve tracer. The measured output current versus input voltage (I_d vs V_{gs}) characteristics were modelled using an equation which incorporates:

1.   A threshold voltage.

2.   A non-linear region up to 5V.

3.   A linear region above 5V.

The measured output current versus output voltage (I_d vs V_{ds}) characteristics were modelled using an equation which incorporates:

1.   Reverse breakdown (due to the parasitic diode formed by the p-type channel and the n-type drain).

2.   A linear region below pinch-off.

3.   A region with a small slope (due to the effective output impedance of the current source).

The static characteristics of the model were measured in a simulated test circuit and are shown in Figure 6.16.

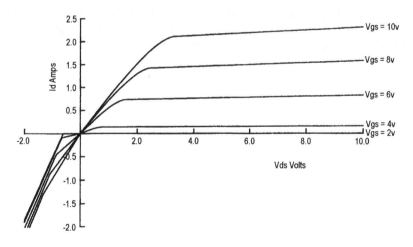

Figure 6.16 DMOS model static characteristics. Graph of I_d vs V_{ds}

The capacitor values for the model were obtained from a.c measurements of the device at 145 MHz. The S parameters of the FET were measured with a network analyser for various gate-source and drain-source voltages. These S parameters were converted to y parameters and the capacitor values were obtained from the y parameters using the following equations:

$$C_{dg} = \frac{-\mathrm{Im}(y_{12})}{\omega} \qquad (6.4)$$

$$C_{ds} = \frac{\mathrm{Im}(y_{22})}{\omega} - C_{dg} \qquad (6.5)$$

$$C_{gs} = \frac{\mathrm{Im}(y_{11})}{\omega} - C_{dg} \qquad (6.6)$$

It should be noted that these equations are approximate as they do not take the lead inductors into account. However, the measurements were performed at low frequencies to reduce the effect of the lead inductances. Further, these equations are only correct as long as the device series resistance is small. The gate source capacitor was found to have an almost constant value of 100pF independent of V_{ds} or V_{gs} however, the other two capacitors were found to be non-linear. The measured and simulated feedback and output capacitors are shown in Figures 6.17 and 6.18 and show close correlation. The drain gate feedback capacitance was assumed to be independent of the drain source voltage and the drain-source output capacitance was assumed to be independent of the drain gate voltage. Measurements indicated that this was a reasonable approximation.

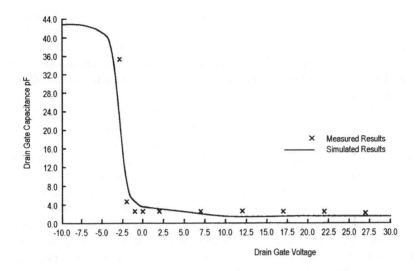

Figure 6.17 DMOS model dynamic characteristics: drain gate capacitance

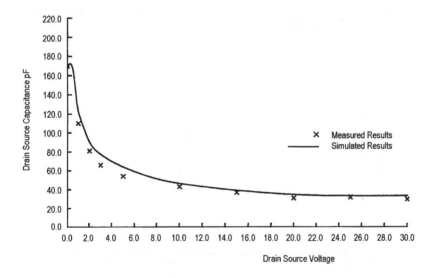

Figure 6.18 DMOS model dynamic characteristics: drainsource capacitance

The capacitance values obtained from the S parameter measurements of the power MOSFET give the small signal change in charge with respect to the voltage across the capacitor. As the total charge in the capacitor needs to be defined as a function of the voltage, it is the integral of the measured curve which is specified in the model. The constant of integration is in the charge on the capacitor when there is no voltage across it and is set to zero. These functions are entered into the model as tables because no suitable equation has been found which would follow the required curves for both positive and negative voltages.

An initial test of the model was made by putting it into a simulated switching circuit with a resistive load and comparing the results with those obtained from an experimental jig. The source of the FET was connected to ground and the drain was connected to a positive 5V power supply via a 9.4Ω resistor. (A 2nH parasitic lead inductance was incorporated in the simulation.) The drain of the FET was connected to a 300ps sampling oscilloscope which presented 50Ω across the drain and source of the FET. The circuit was driven from a 145 MHz sine wave generator with a 50Ω output impedance. A matching network was used at the gate so that the circuit presented a 50Ω impedance to the sine wave generator. The matching network predicted by the model was found to be similar to that required by the power MOSFET when used in a constructed switching circuit. The drain waveform obtained from the simulation (Figure 6.19) is similar to that obtained from the constructed circuit (Figure 6.20). Both of these observations suggest that the model is accurately predicting the FET characteristics under non-linear operating conditions at 145MHz.

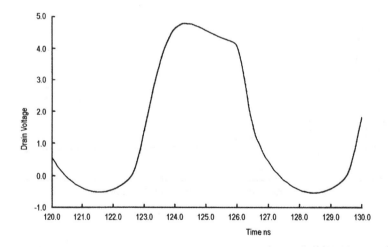

Figure 6.19 MOSFET model switching waveform (V_{supply} = 5V, V_{bias} = 4V, input power = 0.8 W, R_{load} = 9.4//50Ω)

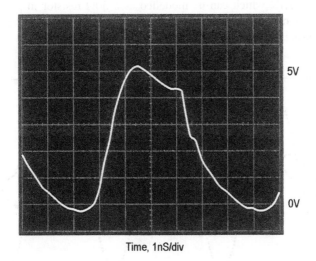

Time, 1nS/div

Figure 6.20 Power MOSFET switching waveform (V_{supply} = 5V, V_{bias} = 4V, input power = 0.8W, R_{load} = 9.4//50Ω), X-axis 1ns/div, Y-axis 1 V/div

6.3.7   CAD of Input Matching Networks

Matching the non-linear impedance of the simulated circuit in the time domain takes a large number of iterations and considerable time; a rapid technique for matching was therefore developed. The drain and the source of the FET were connected to the broadband circuit and the gate was connected to a sinusoidal voltage source set to the correct operating voltage. A CAD Fourier transform was performed to find out the amplitude and phase of the fundamental component of the input voltage and current. The Fourier transform was performed within the CAD package by multiplying the voltage and current by quadrature components and integrating these functions over one period using a current source, switch and capacitor. This ensured high accuracy in the Fourier transform as the CAD package adjusts the time intervals to maintain accuracy. The fundamental impedance was then calculated and a matching network designed. When the network was incorporated into the simulation an almost perfect match was achieved. Figure 6.21 shows the gate voltages and currents at the input of the matching network. It is interesting to note that the gate voltage and the voltage at the input to the matching network are similar in magnitude even though the impedances are very different. This is because the device input impedance was found to be $(1+7j)\Omega$ which can be modelled as a 50Ω resistor in parallel with a capacitor. As the matching network is assumed to be lossless the voltages would be the same to ensure power conservation.

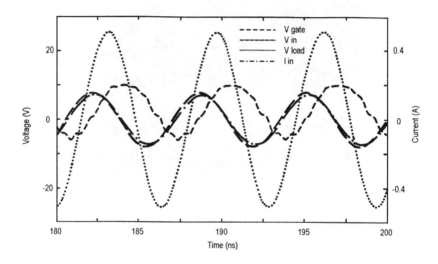

Figure 6.21 Broadband Class E amplifier operating at 155 MHz ($R_{load} = 50\Omega$)

6.3.8 Simulations of the Broadband Amplifiers

Simulations of the complete broadband amplifier were performed with the non-linear model at 155MHz. The input power to the amplifier was 600mW and the output power was 6.25W (figure 6.21). This shows close correlation with the measured results (Figures 6.13 and 6.14). The simulated efficiency at 155MHz using the new non-linear model was 75% and the experimental efficiency was 65%. This shows a lower and more accurate prediction of efficiency than was predicted using the resistive switch model described in Section 6.3.4, the results of which are shown in figure 14. The measured drain-source voltage can be compared with the simulated voltage and both show a peak voltage swing of 35V and a similar shape (Figures 6.22 and 6.23).

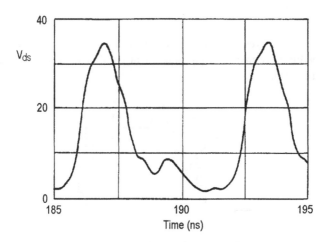

Figure 6.22 Drain-source voltage at 155 MHz, simulated

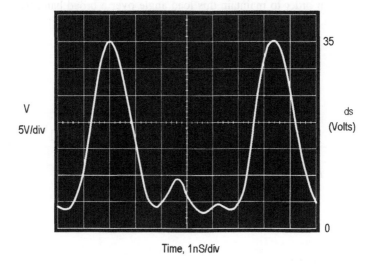

Figure 6.23 Drain-source voltage at 155 MHz, measured

6.3.9   Load Angle Network

A circuit capable of presenting a constant load angle over a broad bandwidth is analysed. The circuit consists of an inductor in parallel with a low Q series tuned circuit (Figure 6.24).

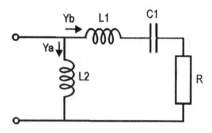

Figure 6.24 Broadband network

The load angle is set by the ratio of the impedance of the inductor L_2 to the resistor R. In order to maintain this load angle over a broad band the slope of the susceptance of L_2 is cancelled by the slope of the resonant circuit L_1, C_1.

$$Y_a = \frac{-j}{\omega L_2} \tag{6.7}$$

$$Y_b = \frac{1}{R + j\left(\omega L_1 - \dfrac{1}{\omega C_1}\right)} = \frac{\omega^2 C_1^2 R - j\omega C_1\left(\omega^2 C_1 L_1 - 1\right)}{\left(\omega C_1 R\right)^2 + \left(\omega^2 C_1 L_1 - 1\right)^2} \tag{6.8}$$

At the resonant frequency of C_1 and L_1:

$$\omega^2 C_1 L_1 - 1 = 0 \tag{6.9}$$

Therefore:

$$Y_b = \frac{1}{R} \tag{6.10}$$

$$Y_a = \frac{-j}{\omega L_2} \tag{6.11}$$

The total admittance is thus:

$$\frac{1}{R} - \frac{j}{\omega L_2} \tag{6.12}$$

For a load angle θ:

$$\tan \theta = \frac{R}{\omega L_2} \tag{6.13}$$

or:

$$L_2 = \frac{R}{\omega \tan \theta} \tag{6.14}$$

The slope of Y_a is:

$$\frac{d\{\mathrm{Im}(Y_a)\}}{d\omega} = \frac{1}{\omega^2 L_2} \tag{6.15}$$

Close to resonance:

$$\left(\omega^2 C_1 L_1 - 1\right)^2 \tag{6.16}$$

tends to zero and the slope of Y_b is:

$$\frac{d\{\mathrm{Im}(Y_b)\}}{d\omega} = \frac{d}{d\omega}\left(\frac{1 - \omega^2 C_1 L_1}{\omega C_1 R^2}\right) = -\frac{\omega^2 C_1 L_1 + 1}{\omega^2 C_1 R_2} \tag{6.17}$$

As:

$$\omega^2 C_1 L_1 = 1 \tag{6.18}$$

$$\frac{d\{\mathrm{Im}(Y_b)\}}{d\omega} = \frac{-2}{\omega^2 C_1 R^2} \tag{6.19}$$

For slope cancellation:

$$\frac{d\{\mathrm{Im}(Y_a)\}}{d\omega} = -\frac{d\{\mathrm{Im}(Y_b)\}}{d\omega} \tag{6.20}$$

and therefore:

$$\frac{1}{\omega^2 L_2} = \frac{2}{\omega^2 C_1 R^2} \tag{6.21}$$

$$C_1 = \frac{2L_2}{R^2} \tag{6.22}$$

For resonance of C_1 and L_1:

$$L_1 = \frac{1}{\omega^2 C_1} \tag{6.23}$$

The design criteria for the network are therefore:

$$L_2 = \frac{R}{\omega \tan \theta} \tag{6.24}$$

$$C_1 = \frac{2L_2}{R^2} \tag{6.25}$$

$$L_1 = \frac{1}{\omega^2 C_1} \tag{6.26}$$

6.4   References and Bibliography

1. R.S. Carson, *High Frequency Amplifiers*, Wiley, 1982.
2. Guillermo Gonzalez, *Microwave Transistor Amplifiers. Analysis and Design*, Prentice Hall, 1984.
3. W.H. Haywood, *Introduction to Radio Frequency*, Prentice Hall, 1982.
4. *Bipolar and MOS Transmitting Transistors*, Philips Application Report.
5. Chris Bowick, *RF Circuit Design*, SAMS, Division of Macmillan, 1982.
6. H.L. Krauss, C.W. Bostian and F.H. Raab, *Solid State Radio Engineering*, Wiley, 1980.
7. N.O. Sokal and A.D. Sokal, *High Efficiency Tuned Switching Power Amplifier*, US Patent No. 3919656, 1975.
8. N.O. Sokal and A.D. Sokal, "Class E - A new class of high efficiency tuned single-ended switching power amplifiers", *IEEE Journal of Solid-state Circuits*, **SC-10**, No. 3, pp. 168–176, 1975.
9. N.O. Sokal, "Class E Can Boost the Efficiency of RF Power Amplifiers", *Electronic Design*, **25**, No. 20, pp. 96–102, 1977.
10. F.H. Raab, Idealised Operation of the Class E Tuned Power Amplifier, *IEEE Transactions on Circuits and Systems*, **CAS-24**, No. 12, pp. 725–735, 1977.
11. F.H. Raab, "Effects of Circuit Variations on the Class E Tuned Power Amplifier", *IEEE Journal of Solid-state Circuits*, **SC-13**, No. 2, pp. 239–247, 1978.
12. A.I. Zverev, *Handbook of Filter Synthesis*, Wiley, 1969.
13. R.A. Minisian, "Power MOSFET Dynamic Large-signal Model", *Proceedings of the IEE*, **130**, Pt I, No. 2, pp. 73–799, 1983.
14. J.K.A. Everard and A.J. King, "Broadband Power Efficient Class E Amplifiers with a Non Linear CAD Model of the Active Device", *Journal of the IERE*, **57**, No. 2, pp. 52–58, 1987.
15. S.C. Cripps, *RF Power Amplifiers for Wireless Communications*, Artech House, 1999.

7

'Real Time' Large Signal Modelling

7.1   Introduction

Modern large signal modelling packages offer extremely accurate results if good models are provided. However, they are often slow when optimisation is required. It would therefore be very useful to be able to optimise the performance of a circuit, such as the load network for a power amplifier, by being able to vary the important parameter values as well as the frequency and then to observe the waveforms on an 'oscilloscope-like' display in real time.

This chapter describes a circuit simulator which uses the mouse with crosshatch and slider controls to vary the component values and frequency at the same time as solving the relevant differential equations. The techniques for entering the differential equations for the circuit are described. These differential equations are computed in difference form and are calculated sequentially and repetitively while the component values and frequency are varied. This is similar to most commercial time domain simulators, but it is shown here that it is relatively easy to write down the equations for fairly simple circuits. This also provides insight into the operation of large signal simulators. The simulator was originally written in QuickBasic for an Apple Macintosh computer as this included full mouse functionality. The version presented here uses Visual Basic Version 6 for a PC and enables the data to be presented in an easily readable format. A version of this program is used here to examine the response of a broadband highly efficient amplifier load network operating around 1-2GHz.

7.2   Simulator

A typical simulator layout is shown in Figure 7.1. It consists of a main 'form' entitled Form 1 which displays four waveforms, and in this case, always shows five cycles independent of frequency to ensure correct triggering. One complete display is then calculated and then updated. The number of display calculations is shown as the number of passes. The simulator allows four variables to be controlled using the position of the mouse in Form 2 and 3 when the left button is pressed. The frequency is varied using a slider control.

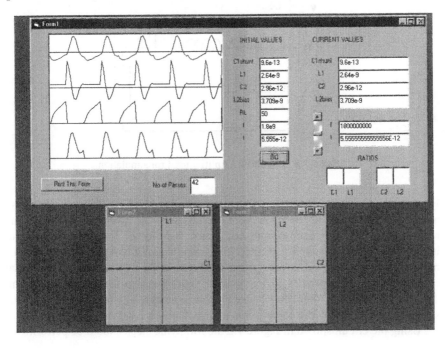

Figure 7.1 Real time large signal circuit simulator displaying waveforms of a broadband amplifier network operating around 1.8GHz. Program written by Peter Turner and Jeremy Everard

The initial and current values of the components are displayed as well as the ratio of the change in value. 500 time steps are calculated per display independent of frequency and 50 points are plotted as most of the processor time for this simulation is taken up in providing the display.

We will now describe how the difference equations can be derived for a circuit by taking an example of the broadband load network shown in Chapter on power amplifiers and shown in Figure 7.2. This circuit looks like a conventional Class E amplifier circuit but the component values are quite different as it is actually optimised for broadband operation with minimal filtering. This simulation technique was originally used to optimise an amplifier circuit containing non-linear step recovery diodes to limit the peak voltage swing at the output.

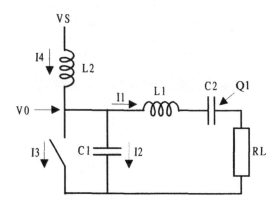

Figure 7.2 Broadband amplifier used for simulation

For ease the circuit is driven by an ideal switch. For rapid analysis, we shall also make approximations about the transient response of the switch on closure. This reduces the transient requirements and hence stability of the software without introducing significant errors. In fact this is a potential advantage of this type of modelling as one can occasionally and deliberately disobey certain fundamental circuit laws for short periods of time without significantly affecting the waveform. The final result can then be checked on a commercial simulator or on this simulator by modelling the components more accurately.

Using the circuit shown in Figure 7.2 the following steps should be performed.

1.   Write down the differential and integral equations for the circuit.

2.   Convert these equations to difference equations so they can be solved iteratively.

3.   Use the mouse and slider controls to control the variation of selected components and frequency.

4.   Plot the required waveforms while showing the values of the varied components.

This will be illustrated for the example shown in figure 7.2. The differential equation for the series arm consisting of L_1, C_2, R_L is therefore:

$$V_0 = L_1 \frac{dI_1}{dt} + \frac{Q_1}{C_2} + I_1 R \tag{7.1}$$

Making the derivative of current the dependant variable:

$$\frac{dI_1}{dt} = \frac{1}{L_1} \left(V_0 - \frac{Q_1}{C_2} - I_1 R \right) \tag{7.2}$$

This is now written in 'difference form' by relating the new value to the previous value. For example:

$$\frac{I_{1(n)} - I_{1(n-1)}}{\Delta t} = \frac{1}{L_1} \left(V_0 - \frac{Q_1}{C_2} - I_1 R \right) \tag{7.3}$$

Therefore:

$$I_{1(n)} = I_{1(n-1)} + \frac{\Delta t}{L_1} \left(V_0 - \frac{Q_1}{C_2} - I_1 R \right) \tag{7.4}$$

Note that in a computer program the $(n-1)$ term can be given the same variable name as the (n) term as the new value assigned to the variable is now equal to the old value plus any changes. In this example the equation written in the program would therefore be:

$$I1 = I1 + \frac{t}{L1} \left(V0 - \frac{Q1}{C2} - I1R \right) \tag{7.5}$$

Note also that the incremental step Δt is now called t, and the subscripts have been removed. Similarly the voltage across C_1 is:

$$V_0 = \frac{1}{C_1} \int I_2 dt \tag{7.6}$$

$$\frac{dV_0}{dt} = \frac{I_2}{C_1}$$

(7.7)

In difference form:

$$\frac{V_{0(n)} - V_{0(n-1)}}{\Delta t} = \frac{I_2}{C_1}$$

(7.8)

As before the form of the equation used in the computer program would be:

$$V0 = V0 + \frac{t\,(I2)}{C1}$$

(7.9)

The voltage across the bias inductor L_2 is:

$$V_S - V_0 = L_2 \frac{dI_4}{dt}$$

(7.10)

therefore:

$$\frac{dI_4}{dt} = \frac{V_S - V_0}{L_2}$$

(7.11)

The difference equation is therefore:

$$\frac{I_{4(n)} - I_{4(n-1)}}{\Delta t} = \frac{V_S - V_0}{L_2}$$

(7.12)

The equation as it would appear in the computer program is therefore:

$$I4 = I4 + \frac{t\,(VS - V0)}{L2}$$

(7.13)

The charge equation for C_2 is:

$$\frac{dQ_1}{dt} = I_1 \tag{7.14}$$

The resulting equation used in the computer program would therefore be:

$$Q1 = Q1 + I1.t \tag{7.15}$$

The current continuity equation is:

$$I3 = I4 - I1 - I2 \tag{7.16}$$

To model the ideal switch it was mentioned that the transient response on closure is deliberately ignored in this case to ease the modelling and reduce the likelihood of software instability. This instability is caused by the huge current spikes on closure of an ideal switch across a capacitor. This switch can also be modelled using a time varying resistor. When the switch is open:

$$I3 = 0 \tag{7.17}$$

When the switch is closed:

$$V0 = 0 \quad \text{and} \quad I2 = 0 \tag{7.18}$$

An example of the part of the Visual Basic computer program used to calculate and plot the solution is shown below. The full code for Forms 1, 2 and 3 is shown in Sections 7.3, 7.4 and 7.5 respectively.

```
'CIRCUIT CALCULATIONS FOR FIVE CYCLES USING 500 POINTS
For X = t To E Step t

k = k + 1
A = CInt((X * f)-Int(X * f))'Switch waveform(5 cycles)
        Q1 = Q1 + (t * I1)
        I1 = I1 + (t * (V0 - (Q1 / C2) - (I1 * RL))/L1)
        I4 = I4 + (t * (Vs - V0) / L2)
        I3 = I4 - I1 - I2
        If A = 1 Then
        I3 = 0
        ElseIf A = 0 Then
        V0 = 0
        I2 = 0
```

```
             End If
             I2 = I4 - I3 - I1
             V0 = V0 + (t * I2 / C1)
             yval1(k) = yval(k)
             yval(k) = 500 - (I1 * 1000)
             yvala1(k) = yvala(k)
             yvala(k) = 1500 - (I2 * 1000)
             yvalb1(k) = yvalb(k)
             yvalb(k) = 2500 - (I3 * 1000)
             yvalc1(k) = yvalc(k)
             yvalc(k) = 3500 - (V0 * 20)
Next X
```

7.3   Form 1 (firstform.frm)

```
Dim n As Integer
Dim p As Long
Dim xval(700) As Integer

Dim yvala(700) As Double
Dim yvala1(700) As Double
Dim yvalb(700) As Double
Dim yvalb1(700) As Double
Dim yvalc(700) As Double
Dim yvalc1(700) As Double

Dim yval1(700) As Double
Dim yval(700) As Double

Private Sub Command1_Click()
Timer1.Enabled = True
End Sub
'GO BUTTON
Private Sub Command2_Click()

    Command2.Visible = False
    Form1.PrintForm
    Command2.Visible = True

End Sub

Private Sub Form_Load()
    Timer1.Enabled = False

    'INITIAL VALUES
    Text4.Text = "9.6e-13" 'C1 series capacitor
```

```
Text5.Text = "2.64e-9" 'L1 series inductor
Text6.Text = "2.96e-12" 'C2
Text7.Text = "3.709e-9" 'L2
Text8.Text = "50" 'Load resistor
Text9.Text = "1.8e9" 'initial frequency
Text10.Text = "5.555e-12"  'initial time step

'CURRENT VALUES
Text11.Text = Text4.Text
Text12.Text = Text5.Text
Text13.Text = Text6.Text
Text14.Text = Text7.Text
Text17.Text = Text9.Text
Text18.Text = Text10.Text

Form2.Visible = True
Form3.Visible = True
f = Val(Text9.Text)
t = Val(Text10.Text)
E = 5 / f
z = 100 * f
L1 = Val(Text5.Text)
C1 = Val(Text4.Text)
C2 = Val(Text6.Text)
RL = 50
L2 = Val(Text7.Text)
Ron = 0.01
Roff = 10000
Vs = 10
Vk = 25
V1 = 0
V0 = 0
I1 = 0
I2 = 0
I3 = 0
I4 = 0
Q1 = 0
VScroll1.Value = 500
p = 0
For k = 1 To 600
    xval(k) = k * 10
Next k


End Sub
```

```
Private Sub Image1_Click()

End Sub
'ENABLE INITIAL VALUES TO BE VARIED BY TYPING ON SCREEN
Private Sub Text10_Change()
    t = Val(Text10.Text)

End Sub

Private Sub Text10_GotFocus()
Timer1.Enabled = False
End Sub

Private Sub Text4_Change()

    Text11.Text = Text4.Text
    C1 = Val(Text4.Text)

End Sub

Private Sub Text4_GotFocus()
 Timer1.Enabled = False
End Sub

Private Sub Text5_Change()
   Text12.Text = Text5.Text
   L1 = Val(Text5.Text)
End Sub

Private Sub Text5_GotFocus()
Timer1.Enabled = False

End Sub

Private Sub Text6_Change()
   Text13.Text = Text6.Text
   C2 = Val(Text6.Text)
End Sub

Private Sub Text6_GotFocus()
Timer1.Enabled = False
End Sub

Private Sub Text7_Change()
   Text14.Text = Text7.Text
   C1 = Val(Text7.Text)
End Sub
```

```
Private Sub Text7_GotFocus()
Timer1.Enabled = False
End Sub

Private Sub Text8_Change()
    RL = Val(Text8.Text)
End Sub

Private Sub Text8_GotFocus()
Timer1.Enabled = False
End Sub

Private Sub Text9_Change()
    f = Val(Text9.Text)

End Sub

Private Sub Text9_GotFocus()
    Timer1.Enabled = False
End Sub

Private Sub Timer1_Timer()
 p = p + 1
k = 0
'DRAW FOUR AXES IN DISPLAY
Picture1.Line (0, 500)-(5595, 500), vbBlack
Picture1.Line (0, 1500)-(5595, 1500), vbBlack
Picture1.Line (0, 2500)-(5595, 2500), vbBlack
Picture1.Line (0, 3500)-(5595, 3500), vbBlack

'CIRCUIT CALCULATIONS FOR FIVE CYCLES USING 500 POINTS
For X = t To E Step t

k = k + 1
        A = CInt((X * f) - Int(X * f))
        Q1 = Q1 + (t * I1)
        I1 = I1 + (t * (V0 - (Q1 / C2) - (I1 * RL)) /
L1)
        I4 = I4 + (t * (Vs - V0) / L2)
        I3 = I4 - I1 - I2
        If A = 1 Then
            I3 = 0
        ElseIf A = 0 Then
            V0 = 0
            I2 = 0
        End If
        I2 = I4 - I3 - I1
```

```
            V0 = V0 + (t * I2 / C1)
            yval1(k) = yval(k)
            yval(k) = 500 - (I1 * 1000)
            yvala1(k) = yvala(k)
            yvala(k) = 1500 - (I2 * 1000)
            yvalb1(k) = yvalb(k)
            yvalb(k) = 2500 - (I3 * 1000)
            yvalc1(k) = yvalc(k)
            yvalc(k) = 3500 - (V0 * 20)
Next X

'Display graph once every 500 calculations every fifth
point


     For m = 1 To 500 Step 5


          'Form1.Picture1.PSet (xval(m), yval1(m)),
vbWhite
          'Form1.Picture1.PSet (xval(m), yval(m)), vbRed
          Form1.Picture1.Line (xval(m), yval1(m))-(xval(m
+ 5), yval1(m + 5)), vbWhite
          Form1.Picture1.Line (xval(m), yval(m))-(xval(m
+ 5), yval(m + 5)), vbRed

          'Form1.Picture1.PSet (xval(m), yvala1(m)),
vbWhite
          'Form1.Picture1.PSet (xval(m), yvala(m)),
vbBlack
          Form1.Picture1.Line (xval(m), yvala1(m))-
(xval(m + 5), yvala1(m + 5)), vbWhite
          Form1.Picture1.Line (xval(m), yvala(m))-(xval(m
+ 5), yvala(m + 5)), vbBlack

          'Form1.Picture1.PSet (xval(m), yvalb1(m)),
vbWhite
          'Form1.Picture1.PSet (xval(m), yvalb(m)),
vbBlue
          Form1.Picture1.Line (xval(m), yvalb1(m))-
(xval(m + 5), yvalb1(m + 5)), vbWhite
          Form1.Picture1.Line (xval(m), yvalb(m))-(xval(m
+ 5), yvalb(m + 5)), vbBlue

          'Form1.Picture1.PSet (xval(m), yvalc1(m)),
vbWhite
```

```
        'Form1.Picture1.PSet (xval(m), yvalc(m)),
vbBlack
        Form1.Picture1.Line (xval(m), yvalc1(m))-
(xval(m + 5), yvalc1(m + 5)), vbWhite
        Form1.Picture1.Line (xval(m), yvalc(m))-(xval(m
+ 5), yvalc(m + 5)), vbBlack
    Next m

    Text3.Text = p
End Sub


Private Sub VScroll1_Change()
'multiply frequency by scroll bar upto +/- 50%
f = ((Text9.Text) * (0.5 + (Format(VScroll1.Value)) /
1000))
Text17.Text = f

'Make 5 cycles and 500 time intervals in one display
independent of frequency
t = 1 / (100 * f)
Text18.Text = t
E = 5 / f

End Sub
```

7.4   Form 2 (secondform.frm)

```
Dim mousepressed As Boolean


Private Sub Form_MouseDown(Button As Integer, Shift As
Integer, Xmouse As Single, Ymouse As Single)
  mousepressed = True
    Ymouse = Ymouse / 1425
        Xmouse = Xmouse / 1560
        C1 = Xmouse * Val(Form1.Text4.Text)
        L1 = Ymouse * Val(Form1.Text5.Text)
        Form1.Text11.Text = Format(C1, scientific)
        Form1.Text12.Text = Format(L1, scientific)
        Form1.Text1.Text = Xmouse
        Form1.Text2.Text = Ymouse
End Sub
```

```
Private Sub Form_MouseMove(Button As Integer, Shift As
Integer, Xmouse As Single, Ymouse As Single)
    If mousepressed Then
        Ymouse = Ymouse / 1425
        Xmouse = Xmouse / 1560
        C1 = Xmouse * Val(Form1.Text4.Text)
        L1 = Ymouse * Val(Form1.Text5.Text)
        Form1.Text11.Text = Format(C1, scientific)
        Form1.Text12.Text = Format(L1, scientific)
        Form1.Text1.Text = Xmouse
        Form1.Text2.Text = Ymouse
    End If
End Sub


Private Sub Form_MouseUp(Button As Integer, Shift As
Integer, Xmouse As Single, Ymouse As Single)
    mousepressed = False
End Sub
```

7.5   Form 3 (thirdform.frm)

```
Dim mousepressed1 As Boolean


Private Sub Form_MouseDown(Button As Integer, Shift As
Integer, Xmouse1 As Single, Ymouse1 As Single)
    mousepressed1 = True
    Ymouse1 = Ymouse1 / 1425
        Xmouse1 = Xmouse1 / 1560
        C2 = Xmouse1 * Val(Form1.Text6.Text)
        L2 = Ymouse1 * Val(Form1.Text7.Text)
        Form1.Text13.Text = Format(C2, scientific)
        Form1.Text14.Text = Format(L2, scientific)
        Form1.Text15.Text = Xmouse1
        Form1.Text16.Text = Ymouse1
End Sub


Private Sub Form_MouseMove(Button As Integer, Shift As
Integer, Xmouse1 As Single, Ymouse1 As Single)
    If mousepressed1 Then
        Ymouse1 = Ymouse1 / 1425
        Xmouse1 = Xmouse1 / 1560
        C2 = Xmouse1 * Val(Form1.Text6.Text)
        L2 = Ymouse1 * Val(Form1.Text7.Text)
```

```
        Form1.Text13.Text = Format(C2, scientific)
        Form1.Text14.Text = Format(L2, scientific)
        Form1.Text15.Text = Xmouse1
        Form1.Text16.Text = Ymouse1
    End If
End Sub

Private Sub Form_MouseUp(Button As Integer, Shift As
Integer, Xmouse As Single, Ymouse As Single)
    mousepressed1 = False
End Sub
```

7.6   Module 1 (Module1.bas)

```
Public f, t, E, z, L1, C1, C2, RL, L2, Ron, Roff, Vs,
Vk, V1, V0, I1, I2, I3, I4, Q1 As Double
Public A As Integer
```

Index

Printed and bound in the UK by
CPI Antony Rowe, Eastbourne

Printed and bound by CPI Group (UK) Ltd, Croydon, CR0 4YY

16/04/2025

14658549-0001